FORSCHUNGSBERICHTE AUS DEM LEHRSTUHL FÜR REGELUNGSSYSTEME

TECHNISCHE UNIVERSITÄT KAISERSLAUTERN

Band 7

T0135540

Forschungsberichte aus dem Lehrstuhl für Regelungssysteme

Technische Universität Kaiserslautern

Band 7

Herausgeber:

Prof. Dr. Steven Liu

Guido Flohr

A contribution to model-based fault diagnosis of electro-pneumatic shift actuators in commercial vehicles

Logos Verlag Berlin

λογος

Forschungsberichte aus dem Lehrstuhl für Regelungssysteme
Technische Universität Kaiserslautern

Herausgegeben von
Univ.-Prof. Dr.-Ing. Steven Liu
Lehrstuhl für Regelungssysteme
Technische Universität Kaiserslautern
Erwin-Schrödinger-Str. 12/332
D-67663 Kaiserslautern
E-Mail: sliu@eit.uni-kl.de

Bibliografische Information der Deutschen Nationalbibliothek

Die Deutsche Nationalbibliothek verzeichnet diese Publikation in der
Deutschen Nationalbibliografie; detaillierte bibliografische Daten sind
im Internet über http://dnb.d-nb.de abrufbar.

ISBN 978-3-8325-3338-0
ISSN 2190-7897

Logos Verlag Berlin GmbH
Comeniushof, Gubener Str. 47,
10243 Berlin
Tel.: +49 (0)30 / 42 85 10 90
Fax: +49 (0)30 / 42 85 10 92
http://www.logos-verlag.de

A contribution to model-based fault diagnosis of electro-pneumatic shift actuators in commercial vehicles

Ein Beitrag zur modellbasierten Fehlerdiagnose von
elektropneumatischen Schaltaktoren in Nutzfahrzeugen

Vom Fachbereich Elektrotechnik und Informationstechnik
der Technischen Universität Kaiserslautern
zur Verleihung des akademischen Grades
Doktor der Ingenieurwissenschaften (Dr.-Ing.)
genehmigte Dissertation von

Guido Flohr
geb. in Berlin

D 386

Tag der mündlichen Prüfung: 12.12.2012

Dekan des Fachbereichs: Prof. Dr.-Ing. Norbert Wehn

Vorsitzender der Prüfungskommission: Prof. Dr.-Ing. Gerhard Huth
1. Berichterstatter: Prof. Dr.-Ing. Steven Liu
2. Berichterstatter: Prof. Dr.-Ing. Steffen Müller

Acknowledgements

The thesis at hand is the result of my work as an external doctoral candidate at the Control Systems Group of the Department of Electrical and Computer Engineering of the Technical University Kaiserslautern and with the Daimler Trucks' Advanced Engineering Department, Stuttgart.

I especially thank Prof. Dr.-Ing. Steven Liu, Head of the Control Systems Group, for exercising his role as my doctoral supervisor with such dedication. Our valuable discussions and his steady encouragement were of great benefit for the advancement of my work.

I very much thank Prof. Dr.-Ing. Steffen Müller, Head of Chair of Mechatronics in Mechanical and Automotive Engineering, for his interest in my thesis and his friendly acceptance as reviewer.

My special thanks goes to Dr.-Ing. Michael Kokes for appointing me to his department and for generously procuring all resources necessary to accomplish my work. His outside the box thinking was a continuous inspiration.

Moreover, I thank all my colleagues at the Group and the Department, in particular Dipl.-Ing. Arno von Querfurth, Dipl.-Ing. Hartmut Raiser, Dipl.-Ing. Shengwei Gong, Dipl.-Ing. Bernd Oeffinger, Dipl.-Ing. Andreas Griesing, Dipl.-Ing. Peter Müller, Dipl.-Ing. Tim Nagel, Dr.-Ing. Daniel Görges, Dipl.-Ing. Stefan Simon, Dipl.-Wirtsch.-Ing. Sven Reimann, M.Sc. Wei Wu, M.Sc. Jianfei Wang, M.Sc.Yun Wan, Dr.-Ing. Philipp Münch, Dr.-Ing. Jens Kroneis, Dipl.-Ing. Michal Izák, Dipl.-Ing. Nadine Stegmann-Drüppel, Dr.-Ing. Martin Pieschel and Priv. Doz. Dr.-Ing. habil. Christian Tuttas for many scientific as well as personal discussions and a congenial time. Thank also goes to the Group's technician Swen Becker and secretary Jutta Lenhardt.

Also, I thank my good friends Dr. Christian Buhrmann and Dr.-Ing. Christoph Prothmann for both their academic and moral support.

I particularly thank my parents and my sister for their confidence and support throughout my academic career.

My deepest gratitude I owe to my fiancée, Judith, for her everlasting encouragement and patience.

Contents

Nomenclature

Acronyms

CAN	Controller Area Network
DML	Decision making logic
ECU	Engine control unit
FDI	Fault detection and isolation
FMEA	Failure mode and effects analysis
LZSS	Lempel-Ziv-Storer-Szymanski compression
MV1	Magnetic valve chamber A
MV2	Magnetic valve chamber B
OBD	On board diagnosis
PCM	Power train control module
PI	Parameter identification
RLE	Run length encoding
SMO	Sliding mode observer
SW	Software
UDS	Unified Diagnostic Service (ISO 14229-1)
ZIP	Format for compressed date

Symbols

δ_f	Error threshold (parameter identification)
$\dot{m}$	Air mass flow
$\hat{x}_{1..4}$	Observed state variables
$\hat{y}$	Estimated output cylinder position
κ	Adiabatic constant for air

$\boldsymbol{F}$	Matrix equation
$\boldsymbol{K}$	Switching gain Matrix
$\boldsymbol{M_s}$	Continuous equivalent vector
$\boldsymbol{Q}$	Design matrix
$\boldsymbol{R}$	Vector of observation errors
$\boldsymbol{S}$	Sliding surface
ν_1	Valve parameter magnetic valve 1
ν_2	Valve parameter magnetic valve 2
ν_{f1}	Orifice parameter fault 1
ν_{f2}	Orifice parameter fault 2
ν_{f3}	Orifice parameter fault 3
$\omega_{1..n}$	Fault induction parameters
ω_c	Rotational speed counter shaft
ω_g	Rotational speed gear box shaft
Φ	Measurement vector
$\Psi_{1..3}$	SMO switching term
ρ	Density
$\sigma_{1..3}$	SMO switching term integral value
Θ	Model parameter
$\Theta_{f1..2}$	Fault parameter
$\xi_{1..3}$	DML thresholds for SMO
A	System matrix
A_1	Piston area chamber A
A_2	Piston area chamber B
A_{v1}	Orifice area magnetic valve 1
A_{v2}	Orifice area magnetic valve 2
A_v	Orifice area of an aperture (general)

B	Run length encoding parameter (compression)
b	Coefficient viscous friction coefficient
c_f	Discharge coefficient
c_p	Specific heat of air at constant pressure
c_v	Specific heat of air at constant volume
$compactiv$	Status compression active
$d(k)$	Differential signal (compression)
$e_{1..4}$	Observation errors
$f(x)$	Nonlinear system function
F_1	Force chamber A
F_2	Force chamber B
F_d	Dynamic friction
F_f	Synchronization counterforce
F_{pt}	Coil spring counterforce
F_p	Force acting on shifting components (general)
F_{rtx}	Force component x direction
F_{rty}	Force component y direction
F_s	Force coil spring
F_s	Static friction
$g(x)$	Nonlinear input function
H	Enthalpy
$h(x)$	Nonlinear output function
h_g	Gauß-Newton step
$J(k)$	Cost function
$J(x)$	Jacobian matrix
k	Index variable (compression)
$K_{1..4}$	Switching gains SMO

$ksize$	Size of compressed data
$ktime$	Compression duration
l	Maximum stroke of shifting cylinder
$L_{1..4}$	Continuous gains SMO
LAB	Number of Bits for compression (compression)
$lenAsvSig$	Length of signal to be compressed
m	Mass of cylinder piston
$M_{1..4}$	Auxiliary matrices
M_f	Torque synchronizer
N	Length of sliding window (compression)
$numBufIdxRd$	Relative index of first memory array
$numBufIdxStrt$	Index of first memory array
$numTrigger$	Cipher for triggered event
p	Pressure (general)
$P_{lenSigShftAbrt}$	offset value (on board)
p_1	Absolute pressure chamber A
p_2	Absolute pressure chamber B
p_i	Physical parameter
P_u	Downstream pressure
P_u	Upstream pressure
$PdDrvSptMax$	Maximum cylinder position (on board)
$PdDrvSptMin$	Minimum cylinder position (on board)
$prssSuplGb8Bit$	Supply pressure (on board)
R	Universal gas constant
r_n	Initial residual (parameter identification)
r_x	Residual for variable x
$r_{y1..3}$	Fault residual (parameter identification)

s	Sliding mode observer term
$s(k)$	Measured input signal (compression)
$sigLen$	Input signal length (compression)
$sign$	Sign function
$sigSelect$	Input signal selector (compression)
$sPosMea$	Cylinder position (on board)
$sSptSns$	Cylinder position (on board)
$stAsvMea$	Control signal magnetic valve (on board)
$stBufDisa$	Trigger freezing ring buffer
$stTrigDisa$	Trigger disabling event identification
T	Absolute temperature
U	Internal energy
u_1	Control input magnetic valve 1
u_1	Control input magnetic valve 2
V	Volume (general)
V_1	Lyapunov function
W	Work
$x_{1..4}$	State variables
y	Sensor output cylinder position
$y(k)$	Compression signal (compression)

Super-Subscript

$\hat{\cdot}$	Estimated values
f_1	Values relating to fault 1
f_2	Values relating to fault 2
f_3	Values relating to fault 3
T	Transpose of a matrix

1 Introduction

1.1 Background and motivation

A high reliability is an essential criterion when buying a modern commercial vehicle and therefore a main development goal (Figure 1.1). Especially the diagnosis of system and component faults is of growing importance. Therewith necessary reliability measures can be taken even in early phases of product development, whereas in later stages of the vehicle lifecycle a fast and precise diagnosis can reduce vehicle downtime and thus costs. However, due to the steadily increasing complexity of these systems and components, computationally extensive diagnosis algorithms are scarcely implementable into electronic control units, updates with newly identified error causes tend to be tardily. Feasible diagnosis concepts have to address the constraints in a practical sense. The most frequent cause for the breakdown of heavy

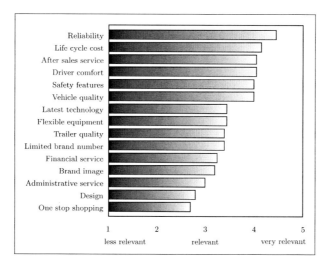

Figure 1.1: Buying criteria for logistics service provider [51]

commercial vehicles are faults of electric and electronic components (Figure 1.2). That is why current diagnosis strategies most generally focus on testing breaks and short-circuit faults of the sensor and component connecting wires. This diagnosis focus neglects pneumatic faults that might eventually lead to breakdowns. To further increase diagnosis depth it is self-evident to introduce more precise measures such as model-based methods to the diagnosis procedure and to apply them to a further range of components that are not exclusively of electric or electronic nature.

Unlike in passenger cars shifting automation in heavy commercial vehicles is not realized by hydrodynamic torque converters but electro-pneumatic shift actuators in combination with automated central clutches. Because of the non-linearity that adheres to pneumatic system due to the compressibility of air, well established diagnosis techniques for this class of systems are rather seldom in practical applications. A further constraint when implementing pneumatic models in automotive environments results from the reduced number of pressure sensors employed due to their high cost.

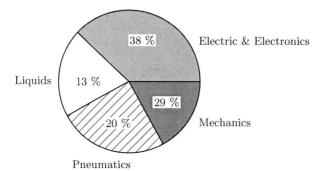

Figure 1.2: Breakdown causes of heavy commercial vehicles [43]

Introductory state of the art review

In the last two decades model-based diagnosis has been addressed by a large number of publications throughout several technical disciplines. This introductory section gives a brief insight to diagnosis concepts currently introduced into the automotive industry. A more detailed state-of-the-art review will be presented at the beginning of each subsequent chapter.

A comprehensive insight to model-based monitoring and fault diagnosis for automotive applications is given by Isermann [39], clearly explaining the fundamental concepts and methods. Qin et al. [64] present a fault diagnosis strategy for an automated manual transmission based on analytical redundancy that is designed to totally reside within the electronic control unit. A linearized gear box model is employed to detect and isolate mainly sensor faults and indicate them by error codes.

Assaf and Dugan [5] combine an onboard fault tree model with an additional sensor layer to predict and report component failure prior to a possible vehicle breakdown. This approach, that combines heuristic knowledge with signal based principles, is applied to an automotive cooling and heating system.

A remote diagnostic approach is introduced by Carr [12]. A client equipment within the vehicle is equipped to send relevant diagnosis data to a distant server via wireless connections. Different advanced diagnostic capabilities in terms of a remote data acquisition are denoted but not further elaborated. You et al. [81] analyze the current status of vehicle remote diagnostics and maintenance and their future potential as well as customer benefits. An exemplary solution is presented on a conceptual level. Fritsch et al. [31] introduce a novel scheme of a model-based remote diagnosis by decomposing the diagnostic task into on-board as well as off-board components under performance considerations. The on-board component exclusively solves the problem of fault detection, fault isolation and identification is performed off-board. The systems to be diagnosed is modeled as a discrete event system.

The concept of utilizing a symptoms-based fault detection and data acquisition unit on-board and a model-based fault isolation scheme off-board is the foundation of this thesis. The basis of this concept has been previously presented [28][43].

1.2 Thesis outline and contribution

Motivated by the findings above the aim of this thesis is to design and implement a distributed fault diagnosis concept for an electro-pneumatic shift actuator under practical restrictions. One of the fundamental restrictions is the limited access to a commercial vehicle's ECUs due to safety considerations. Furthermore, as of cost requirements, only a limited numbers of sensors is present in the vehicle. The central question answered by the thesis at hand is wheather or not it is possible to perform an effective diagnosis under these practical restrictions. The thesis' contributions for solving the task will be:

- Definition of a new diagnosis concept with distributed components

- Implementation of a symptoms-based fault detection and data acquisition unit

- Design of a model of an electro-pneumatic shift actuator

- Performing a parameter identification for the developed model

- Design of a fault diagnosis scheme based on nonlinear parameter identification

- Design of a fault diagnosis scheme based on a sliding mode observer

- Setup of an automated manual transmission test bench

- Performing realistic measurements to verify the fault diagnosis schemes

The new diagnosis concept with distributed components will be presented in Chapter 2. The concept is embodied in an on-board unit holding the symptoms-based fault detection and data acquisition unit, while fault diagnosis is performed in a workshop tester. All relevant components of the concept and the setup of the overall diagnosis chain will be described in detail.

Chapter 3 begins with a short literature review of pneumatic modeling techniques that have been established over the past decades. In the following a physical model of the electro-pneumatic shift actuator for an automated manual transmission is derived. The model is identified and validated experimentally using a number of test bench measurements. Some model simplifications are performed and thoroughly discussed. In a further section common faults of pneumatic actuators are characterized and a subset of them modeled for diagnostic purposes. This results achieved in this chapter are the foundation for both model-based fault diagnosis schemes discussed in the next chapters.

As one of the two diagnosis schemes that are part of this thesis a fault diagnosis based on nonlinear parameter identification is scrutinized in Chapter 4. A preliminary section discloses the evolution from system identification methods to fault diagnosis techniques and the physical interpretation required in doing so. The problem of residual generation using algorithm of Levenberg-Marquardt is formulated, the necessary conditions such as fault detectability and isolability is discussed. Again simulations are performed using test bench measurement data.

In Chapter 5 a fault diagnosis scheme based on the evaluation of a sliding mode observer's sliding term is constructed. The observer is designed such that it maintains a stable sliding motion in the presence of a fault even though discontinuities in the state space are present. Prior to that some observability measures for nonlinear dynamic systems are introduced, the observability analysis of the pneumatic actuator is performed. In the review section of this chapter different observer based fault isolation methods are discussed, namely dedicated and generalized schemes. The chapter is concluded by proper simulation preparing the next chapter.

For Chapter 6 all fault scenarios that have been introduced in Chapter 3 are implemented and thoroughly tested on a realistic test bench environment. The detailed results are presented and subjected to a conclusive discussion.

Finally Chapter 7 draws the conclusion and gives recommendations for future work.

2 A novel diagnosis concept with distributed components

2.1 Introduction

In the very beginning of this work the idea of separating diagnostic functionalities and strictly assigning them to either on-board or off-board components was merely a vague idea driven primarily by practical requirements. As denoted in Chapter 1 these challenges – technological as well as procedural – apply when designing a diagnosis concept in the automotive environment. Therefore the diagnosis related requirements, benefits and constraints are discussed along desired properties of fault diagnosis systems at first.

A direct comparison between the prevailing state-of-the-art diagnosis concept that mainly originates from OBD (on board diagnosis) regulations and the novel distributed concept with a decomposed diagnosis chain shows the benefits of the latter one. In Section 2.2 the newly introduced concept is specified more precisely. In Section 2.3 the on-board symptoms-based fault detection and data acquisition unit are described in more detail. It is comprised not only by a simple diagnostic model but a modular ring buffer, a signal preprocessing unit and a compact compression algorithm. The off-board components that are involved in this decomposed diagnosis chain, namely the decompression algorithm and the fault diagnosis schemes that operate on either the basis of parameter estimation or state observation techniques, will be presented in the later Chapters 4 and 5.

2.1.1 Desired properties of fault diagnosis systems

A fault diagnosis system shall have certain general properties, the most desirable ones being the following [71]:

(1) **Early detection and diagnosis:** The diagnosis system shall be able to detect and isolate faults or faulty behavior prior to a possible vehicle breakdown. This means that the diagnosis system should be sensitive to incipient faults while not causing false alarms.

(2) **Isolability:** Is the capability of a diagnostic systems not only to detect a fault but to locate it clearly. Isolability of a fault does not only depend on the way the faults affect the system output but also the diagnosis system design.

(3) **Robustness:** In practical settings it is naturally impossible to avoid uncertainties. Therefore it is essential to guarantee at least some robustness to measurement noise, system disturbances and modeling errors.

(4) **Detectability of new faults:** The standard industry measure when designing a diagnosis system are failure mode and effects analysis (FMEA). Nevertheless possible fault and fault tendencies might have been neglected. It is therefore expected from a diagnosis system to cope with these unexpected faults, not wrongly classifying them.

(5) **Accessibility:** As soon as a new fault is noted, the diagnosis system has to be updated within a certain time frame to maintain it's effective mode of operation. Commercial vehicles for example undergo a periodic maintenance only annually.

(6) **Reasonable storage and computational requirements:** Memory and computational requirement are the two fundamental characteristics of diagnosis algorithms that have to be implemented in electronic control units. It can be generally assumed that the diagnostic functionality is not the task with the highest computational priority and has to subordinate for instance to control algorithms.

(7) **Modularity:** Since the responsibility for the development of diagnostic functionalities may well lie in different engineering departments, the development schedules may vary significantly. An applicable diagnosis system has to come with a certain degree of modularity to be able to compensate for that.

2.1.2 Design constraints for a practical implementation

Figure 2.1 shows both diagnosis concepts on a conceptual level. The upper half of the figure shows the state-of-the-art diagnosis concept, where all model-based diagnosis functionalities reside exclusively in the on-board unit. The output information is merely a fault code that indicated the occurrence of an fault related event and is the only input information for the workshop testing unit on the right hand. The lower half depicts the novel distributed concept with a decomposed diagnosis chain. The on-board unit holds the symptoms-based fault detection and data acquisition components whereas the model-based fault isolation is performed in the workshop tester. Environmental data is available for extended analysis. Both diagnosis concepts were checked against the desired criteria of fault diagnosis systems introduced in the section. The results are presented in Table 2.1, wherein the desired properties from Section 2.1.1 are listed in the first column. Clearly can be seen that the advantages of the novel distributed concept with a decomposed diagnosis chain are preponderant. The concept shall be applied to an electro-pneumatic shift actuator

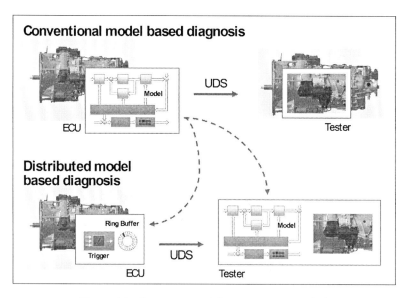

Figure 2.1: Comparison of diagnosis concepts [43]

of automated manual transmission that will be described in the next section.

Desired property	Strict on-board diagnosis	Distributed diagnosis
(1)	++	O
(2)	O	+
(3)	+	+
(4)	+	O
(5)	O	+
(6)	O	++
(7)	-	+

Table 2.1: Diagnosis concepts comparison table

2.1.3 Description of the test object

The sample application of the concept proposed in this thesis will be the functional diagnosis of an electro-pneumatic shift actuator that is part of an automated manual transmission commonly used in heavy commercial vehicles. The main reason for that is that the electro-pneumatic shift actuators were identified to be subject to a high amount of deterioration with the need for further tracking. Figure 2.2 shows an automated manual transmission and an electro-pneumatic shift actuator.

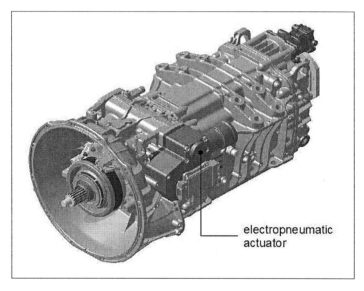

electropneumatic
actuator

Figure 2.2: Transmission for heavy commercial vehicles [24]

2.2 Overall diagnosis chain

As already analyzed and shown in Table 2.1 the implementation of a model-based diagnosis in the on-board electronic control unit cannot be done in full scale due to high computational requirements. This is especially true for the later introduced parameter estimation techniques which would highly exceed the cost-benefit ratio. The on-board storage requirement can only be met by significantly reducing the amount of data required for diagnosis. To meet the overall requirements adequately the previously proposed distribution of the diagnosis components is elaborated in detail. Whereas the previous section introduced the concept of the distribution of diagnosis components on a very conceptual level, the following section describes the approach along a concrete example.

Decomposition of the diagnostic task

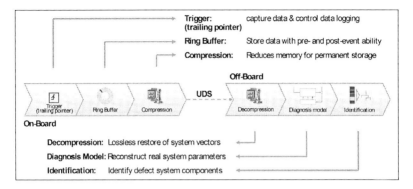

Figure 2.3: Distributed diagnosis chain [43]

Figure 2.3 depicts the decomposition of the diagnostic task into on-board and off-board units, both being composed of three sub-components that are listed below.

On-board components

- **Symptoms-based fault detection unit:** In comparison to existing approaches the on-board data unit will be strictly signal based, reducing computational costs even further and allowing to capture sporadic errors.

- **Ring buffer:** The ring buffer employs a cyclic memory array for storing sensor and control signals for a predetermined time. Once the symptoms-based fault detection unit detects a faulty event the ring buffer is stopped.

- **Compression algorithm:** The function of the compression algorithm is to losslessly compress data from the ring buffer into a structured string to be saved in the non-volatile memory of the on-board electronic control unit.

Data transfer

- **Data transfer connection:** The transmission of the data will be established via a 'static' network. The sensor and control signals stored by the on-board data acquisition unit are transferred in the workshop by a wired connection using unified diagnostic service.

Off-board components

- **Decompression algorithm:** The function of the decompression algorithm is to reconstruct sensor and control signals from a structured string saved in the non-volatile memory of the on-board electronic control unit.

- **Model-based fault isolation unit:** The task of the fault isolation unit is to isolate a certain fault within the diagnosed system by analyzing sensor and control signals using model-based techniques.

- **Fault visualization:** The translation of the isolated faults into practical fault-tree-based repairing instructions is performed in the fault identification and visualization unit. The procedure is straight forward as a palpable decision is required.

2.3 Vehicle on-board fault detection and data acquisition

The task of the symptoms-based on-board fault detection and data acquisition is the automatic detection of faulty events and the storage of relevant sensor, control and ambient data prior to it. To serve that purpose a ring buffer has to charge the above mentioned data chronologically at all times. Once a faulty event is detected the ring buffer is stopped, holding the fault related data in its memory. Following that, a fault specific subset of the ring buffer data is selected and efficiently compressed using either an adapted LZSS [1] algorithm for sensor data or run-length encoding for pulse width modulated input data.

A little later in this section signal-based fault detection is reviewed and some definitions are introduced. All on-board components are described separately. For now the overall on-board functionality shall be explained according to Figure 2.4. The live date is acquired from the data interface of the electronic control unit and fed into the trigger block as well as the preprocessing for the ring buffer. Once an event is detected the ring buffer is stopped and data is passed to the compression unit which disables the trigger block for the time required for compression.

---

[1]the original Lempel-Ziv-Storer-Szymanski algorithm was created in 1982 by James Storer and Thomas Szymanski

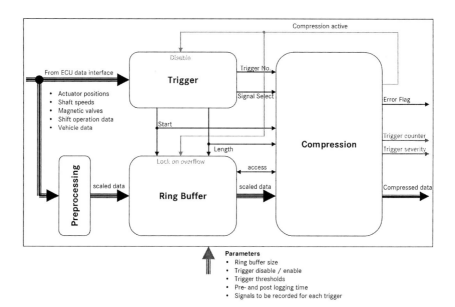

Figure 2.4: Functional block diagram on-board components

2.3.1 Review of signal-based fault detection

Signal-based fault detection is based on the direct analysis of measured signals without the use of process models. As an indicator for an occurred fault the following methods amongst others can be used:

- instantaneous value $y(t)$ for monitoring the absolute value

- the derivative of the measured signal $\dot{y}(t)$ for monitoring the trend

- parameter of the transformed signal i.e. Fourier analysis

It is to be tested whether or not the measured signal remains within a certain range of tolerance. The boundaries have to be designed such that an adequate distance towards a critical value remains while false alarm are circumvented. Signal-based fault detection techniques can be applied straight-forwardly provided that an indicator is found that can be mapped to a fault distinctly. In terms of system excitation, two diagnosis methods can be distinguished [27]:

- **Active diagnosis:** the signals required for diagnosis are derived from an explicit excitation of the process by either step or ramp functions, pseudo noise signals or sinusoidal functions. The parameters of the excitation functions have to be chosen such that all system modes are represented. When active diagnosis is applied it has to be guaranteed that the process itself is not negatively affected or disturbed. Excitation functions can be standardized.

- **Passive diagnosis:** the signals used for diagnosis are solely the ones from the control of the process itself. In contrast to an active diagnosis the signal monitoring unit has to be enabled permanently as soon as the process is started. The advantage is the autonomous functioning and the fact that the process to be controlled remains totally unaffected. Nevertheless excitation in this context is subject to changes depending on the control situation.

2.3.2 Symptoms-based fault detection unit

In this work the diagnosis system is of passive nature because of the implementation in a safety relevant environment. In this case the record of the diagnosis input signals has to be triggered automatically by if a certain threshold is exceeded. The length of the recording shall be at least the length of the impulse response [27].

The symptoms-based fault detection unit visualized in Figure 2.5 serves as the central module of the whole on-board unit containing its *intelligence* and controlling the whole application flow shown in Figure 2.4. The task of the unit is to detect a fault or faulty behavior related to the transmission and consists of two major functional blocks *event identification* and *state control*. The latter coordinates the fault detection and the ring buffer such that the identification of events and the storing of the relevant data are carried out properly. The signal *stTrigDisa* is used for deactivating the event identification, *stBufDisa* for freezing the data in the ring buffer and therefore putting it on hold.

The *event identification* block itself is designed to match certain faulty behaviors such as *unintended gear changes* or *exceeded shifting durations* that both can be caused by shifting components faults such as leakages or increased friction. Once an event that relates to a faulty behavior is detected, the block outputs information namely the event classification number *numTrigger*, the first index *numBufIdxStrt* of the ring buffer array that holds the relevant data connected to the event and the length of this very signal to be compressed *lenAsvSig*. Two examples of how these events can be tackled using symptoms-based intelligence shall be presented now.

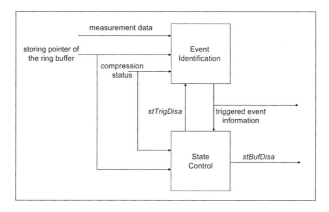

Figure 2.5: Functional block diagram trigger

Unintended gear change

This trigger is characterized by the movement of an electro-pneumatic shift actuator without any activation of magnetic valves. Therefore the trigger is active if a gear is engaged and the clutch is closed. The schematics can be seen in Figure 2.6. Inputs for the block are furthermore the current cylinder position *sSptSns*, the cylinder piston positions limits *PdDrvSptMin* and *PdDrvSptMax* and a safety threshold value. Figure 2.7 shows how the current sensed cylinder position is compared to the engaged/disengaged areas defined by the piston position limits. Once the sensed position of the pneumatic actuator leaves the engaged position an integrator starts summing up the difference between the actual measured position and the last sensed position which was identified as engaged resulting in y_{k+1}:

$$y_{k+1} = y_k + abs(PdDrvSptMin - sSptSns) \tag{2.1}$$

If y_{k+1} exceeds the predefined threshold the trigger is activated.

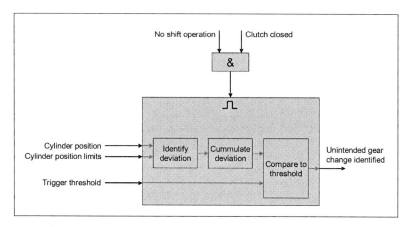

Figure 2.6: Application flow for symptom unintended gearchange

Exceeded shifting duration

The functional diagram of the symptom *exceeded shifting duration* is given in Figure 2.8. Here the length of the time interval that is required by an electro-pneumatic shift actuator to move its position from engaged to disengaged position or vice versa is determined. Input information for the functional block is the actuator position, the magnetic valve status and the shifting request status. The symptom is detected once the sensor signal exits the diagnostic window spanned by engagement positions and maximum allowed shift duration threshold.

2.3.3 Cyclic pre-storage of data

The ring buffer is used for storing sensor and control signals such as the activation of magnetic valves *stAsvMea*, the position information of the sensors *sPosMea* and

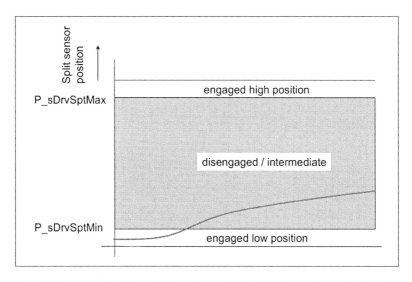

Figure 2.7: Detection of an unintended gearchange of the shift actuator

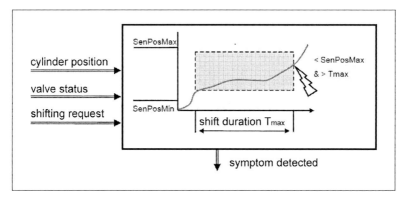

Figure 2.8: Detection of an exceeded shifting duration of the shift actuator

the pneumatic supply pressure *prssSuplGb8Bit*. In Figure 2.9 the ring buffer is visualized with indices n and two pointers. The store pointer indicates where the newly measured data is stored whereas the read pointer indicates the location of the current data held in the buffer. The interlock of these memory areas is guaranteed automatically. In Figure 2.10 the functionality of the cyclic memory is explained

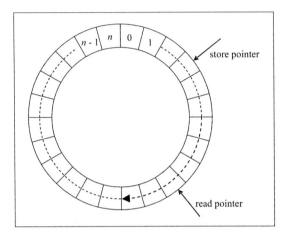

Figure 2.9: Ringbuffer for storing sensor and control signals

for the example from the previous section. *Pre-trigger* and *post-trigger* represent additional time-slots to ensure the storage of the overall shift operation. The *store-pointer* that marks the instantaneous memory index is calculated by subtracting the pre-trigger index from the actual index at the beginning of the shift operation and adding the post-trigger index afterwards. The *read-pointer* indicated the very beginning of the memory slot required for further off-board analysis. The overall

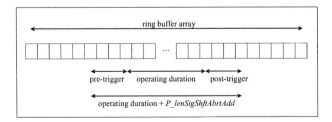

Figure 2.10: Exemplary ringbuffer storage arrangement

ring buffer functionality is depicted in Figure 2.11. An exact memory mapping is performed for reading the ring buffer data off board. The input *stBufDisa* is imple-

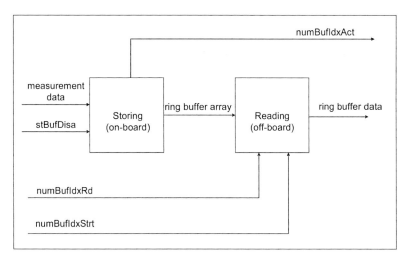

Figure 2.11: Functional block diagram ring buffer

mented to prevent the ring buffer from storing new data and therewith overwriting fault relevant ones. The input *numBufIdxRd* is the array index of the ring buffer relative to the start index pointed by the input *numBufIdxStrt*, the sum of them being equivalent to the *read-pointer* introduced above. *numBufIdxStrt* marks the above *store-pointer*. In the current design the ring buffer holds an overall time interval of *3,6 seconds*.

2.3.4 Data reduction

The general function of the compression algorithm is to losslessly compress the data from the ring buffer into a structured string to be saved in the non-volatile memory of the on-board electronic control unit. The compression algorithm is provided with the information about which signals from the ring buffer have to be compressed. The concrete data to be compressed is determined by signal selection, signal length, signal start index information (Figure 2.12). The compression is set active upon an enable flag and reports compression status, time that was required to compress the signals, size of the compressed string and errors that occurred during compression. The compression is implemented as handwritten C++ libraries [2] and integrated into the *Simulink environment* via s-function builder. The compressed string itself is comprised of a varying selection of signals that are a subset from those held in the

---

[2]The libraries were implemented in cooperation with the Control Systems Group TU Kaiserslautern

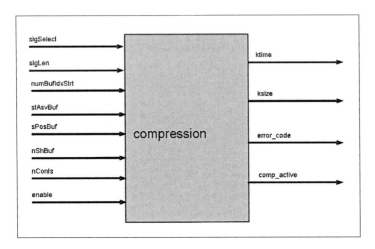

Figure 2.12: Implementation of onboard compression

ring buffer and a compression header as depicted in Figure 2.13. As it is required

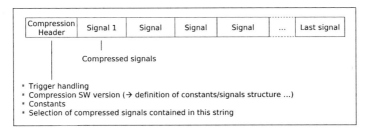

Figure 2.13: Structure of a typical compression string

for the integration of the compressed string into the memory management of the on-board electronic control unit the header has to be provided with a certain structure. Figure 2.14 shows the first two bytes containing the trigger number and significance of the stored event. The compression version is linked to the further setup of the header and the structure the signals are stored in. Furthermore the compression header contains the signal selection information and all constants which are to be stored with the data. All sensor and control signals that are marked with a *high bit* by the signal selection logic will be stored in the sequence. As an example Figure 2.15 visualizes the compression sequence for 28 selected signals. The compression starts upon the *enable* flag. The flag *comp active* indicates that a compression is in progress. The compression algorithm operates in 3 stages which shall be briefly

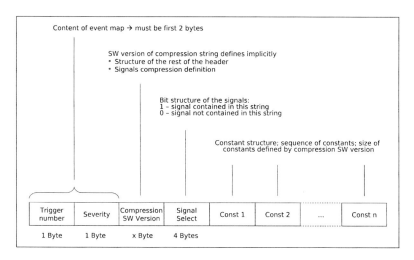

Figure 2.14: Structure of a typical compression header

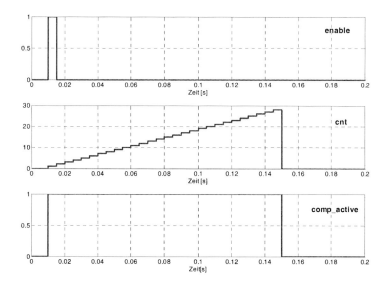

Figure 2.15: Compression sequence

discussed according to Figure 2.16:

Figure 2.16: The three stages of compression

Data preprocessing

The compression signal $y(k)$ is calculated from measured input signal $s(k)$ using the following steps:

- for every $k \geq 1$ calculate the differential signal $d(k) = s(k) - s(k-1)$

- transform $y(k)$ to positive numbers: $y(k) = 2d(k) \; for \; d(k) >= 0$
  $y(k) = -2d(k) - 1 \; for \; d(k) < 0$

In doing so signal entropy[3] $H = \sum_i p_i \times log_2(p_i)$ can be reduced, wherein p_i is the probability of a random symbol.

Compression stage

The compression algorithm applied in this work is *LZSS* as a modification of *LZ77* which is widely known as the algorithm behind *ZIP* compression. *LZSS* is defined for real positive signals $y(k)$. As can be seen from Figure 2.17 the algorithm uses a two parted sliding window that is moved over signal $y(k)$. The preview-buffer to the right holds the elements which are to be compressed, whereas the dictionary on the left holds the elements that have been compressed already. In every compression step the dictionary is searched to fit a maximum correlation with the elements from the preview-buffer. This correlation is being saved as *[0,offset,correlation length]* triple. The offset in the dictionary is the distance from the first element of the correlation to element x. The first number is a flag that gives information about how to interpret the triple. LAB denotes the number of bits needed to code the offset. Overall length of the sliding window N can be calculated as:

$$N = \left(2^{(LAB+1)}\right) + 2 \qquad (2.2)$$

---

[3]in information theory entropy is a measure of for the mean information content of one bit per source, i.e. the mean number of bits required to clearly isolate two symbols from each other

If there is no correlation between preview-buffer and dictionary the element x is coded as $[1, 0, x]$. The LZSS compression algorithm generates triple that represent a signal $y(k)$ losslessly.

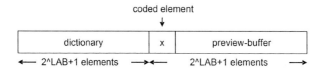

Figure 2.17: Basic principle of LZSS compression

Post processing stage

The triple generated in the *compression stage* holds compressed information byte wise. To efficiently reduce memory usage the information has to be changed to bitwise coded which requires a fixed set of rules. The method used is unary coding where the correlation is coded using the following look up table: The triple's new

length of correlation	code representation
1	1
2	01
3	001
4	0001
...	...

Figure 2.18: Code representation according to LZSS algorithm

representation is *[1, XX...XX, 00...01]* where the first *1* is the flag, *XX...XX* is the binary coded offset and *00...01* a representation of the length of the correlation according to the Table 2.18. The procedure demonstrated above is applied to signals represented in real numbers. *Run-length-encoding* is used instead for binary signals that hold a constant value for a longer period of time. This method is applicable for control input of magnetic valves as there are only a few impulses compared to the overall length of the ring buffer sequence. For these types of signals *LZSS* would work rather inefficiently. The overall compression is quite simple as it only requires one operating stage.

Compression stage run-length-encoding

Run-length-encoding counts for how long a signal is repeated without interruption. The length of this sequence is saved. In addition a parameter B is used to encode the number of bits required to represent the sequence. Maximum run-length is therefore limited to $2^B - 1$. Every element will be encoded with $B + 1$ the tuple structure is simply *[the bit itself, run-length]*. Figure 2.19 shows an example for a given signal to be coded in *RLE*. The parameter B is chosen to be 0, the signal starting bit is 0 at position 0 in byte 0.

	decimal	binary							
Byte		bit 0...bit 7							
0	0	0	0	0	0	0	0	0	0
1	127	0	1	1	1	1	1	1	1
2	136	1	0	0	0	1	0	0	0
3	31	0	0	0	1	1	1	1	1
4	224	1	1	1	0	0	0	0	0

Figure 2.19: Code representation RLE

3 Modeling and identification of an electro-pneumatic shift actuator

3.1 Introduction

3.1.1 State-of-the art review

Pneumatic actuators have been introduced to a variety of applications across industries ranging from industrial handling to active suspensions, assisted breaking and eventually automated shifting. They inhibit a number of favorable attributes such as low specific weight, high power rate as well as clean and safe operation at a relatively low cost. The band of the actuating force ranges from medium to high for medium to large strokes [40]. Originally just operated in bang-bang control, more sophisticated tasks such as exact positioning came into scope with the rising availability of computational capabilities. That is why in recent years research on pneumatic systems has been gaining more and more momentum mainly from a control point of view. Nevertheless a solid foundation of modeling pneumatic actuators and components has been in existence for some decades now. Ali et al. [4] gives a good state-of-the art review. Further notable recent papers on pneumatic modeling and control were published by: Swarnakar et al. [72] Pandian et al. [61] Ilchmanm et al. [36] Rincón-Pasaye and Martínez-Guerra [67] Schwarte [69] Richer and Hurmuzlu [65] Richer and Hurmuzlu [66] Saif and Xiong [68].

Modeling of pneumatic actuators though turns out to be quite sophisticated, as it requires theoretical analysis combining thermodynamics, fluid dynamics and mechanics. The compressibility of air and the highly nonlinear flow through the commanding valves add up to complexity. Generally speaking three modeling steps are required, namely the determination of the mass flow through the orifice of the valve, calculation of the cylinder air properties temperature, pressure, volume and the determination of the load dynamics. Identification techniques are employed to substantiate the mathematical model.

3.1.2 Overview of pneumatic shift components

The electro-pneumatic shift actuator depicted in Figure 3.1 and to be discussed intensively, is an integral part of an automated shifting system utilized by modern automated manual transmissions. Being comprised of two 3/2-way PWM-controlled magnetic valves driving a differential pneumatic cylinder with a shifting assembly connected to the cylinder rod, the position of the actuator is measured by an inductive sensor connected to the cylinder rod directly. In this chapter a complete as well

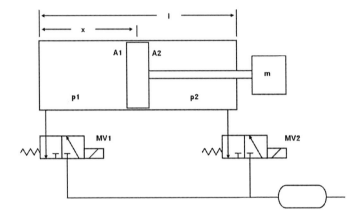

Figure 3.1: Differential pneumatic cylinder and magnetic valves

as a simplified physical model will be derived. The first model is intended to fully understand the effects of the overall system and its interactions and will be the basis for the parameter estimation based fault isolation scheme described in Chapter 4. From that a simplified model is derived as a basis for the observer-based technique. Some feasible assumptions are made. The theoretical analysis and modeling can be split up into a mechanical and a pneumatic subsystem. The latter is comprised of a subsystem for the differential pneumatic cylinder and for the magnetic valves.

3.2 Physical system modeling

To fully understand the complex behavior of the electro-pneumatic shift actuator it is clear that all physical effects and environmental interactions of the system have to be scrutinized in the beginning. As one moves on to the system identification and even further to computational implementation model simplification is inevitable but admissible. This shall be discussed along the following sections.

3.2.1 Shifting-assembly dynamics

On the topmost level Newton's second law is applied to describe the system dynamics:

$$m\ddot{x} + b\dot{x} + F_f + F_l = p_1 A_1 - p_2 A_2 \tag{3.1}$$

where m is combined mass of piston, rod and external load mass, b is the coefficient for the viscous friction, F_f the force of the coulomb friction and F_l the load force. p_1 and p_2 on the right hand side of Equation (3.1) denote the absolute pressure in the cylinder chambers, A_1 and A_2 the effective piston areas respectively. Along Equation (3.1) the system is further detailed in the next steps beginning with the differential pneumatic cylinder as the force-exerting component.

3.2.2 Differential pneumatic cylinder

In this section the relationship between the right hand side of Equation (3.1) with the mass flow rate into and out of the cylinder chambers is derived. In contrast to hydraulic systems the compressibility of air is the dominating factor in pneumatic systems accounting for their highly nonlinear behavior. In modeling practice the gas, which is atmospheric air in most cases, is assumed to be an ideal gas with a homogeneous distribution of temperature and pressure. Furthermore kinetic and potential energy properties are neglected. A closed volume of gas can therefore be described by the ideal gas law,

$$pV = mRT \tag{3.2}$$

where p is the absolute pressure, V the volume, m the mass, R the universal gas constant and T the absolute temperature. If there is mass entering or leaving the volume, the conservation of mass condition must hold. The mass flow rate writes out as

$$\dot{m} = \frac{d}{dt} m = \frac{d}{dt} \rho V \tag{3.3}$$

likewise

$$\dot{m}_{in} - \dot{m}_{out} = \dot{\rho} V + \rho \dot{V} \tag{3.4}$$

The volume can be described by the change of internal energy which results from the addition or loss of enthalpy and work to each pneumatic chamber:

$$\dot{U} = \dot{H} - \dot{W} \tag{3.5}$$

where U is the internal energy, H the enthalpy and W the work done. The time rate of change of the internal energy $\dot{U}$ again can be expressed as:

$$\dot{U} = \frac{d}{dt} \left[m c_v \left(T_s - T_{in} \right) \right] = \frac{d}{dt} \left[m c_v T \right] = \frac{d}{dt} \left[\rho V c_v T \right] \tag{3.6}$$

where mass, temperature, volume and density of the air in each chamber are represented by m, T, V and ρ and T_s is a reference temperature for the calculation of the internal energy. T_i denotes the absolute temperature of the gas flow entering a chamber. c_v is known as the specific heat of air at constant volume. Together with c_p as the specific heat of air at constant pressure, Equation (3.2) for the ideal gas can be rewritten as:

$$\rho T = \frac{p}{R} = \frac{p}{c_p - c_v} \tag{3.7}$$

Hence, Equation (3.6) reformulates to:

$$\dot{U} = \frac{d}{dt}\left(\frac{c_p p V}{c_p - c_v}\right) \tag{3.8}$$

Remembering the right hand side of (3.5) in the change of enthalpy can be described by:

$$\dot{H} = \dot{m}_{in} c_p T_{in} - \dot{m}_{out} c_p T \tag{3.9}$$

assuming that the gas entering the chamber is already of the same temperature as the gas inside, i.e. the enthalpy change due to rate of change of temperature is relatively small compared to the change as an account of the mass flow rate, the enthalpy change can be rewritten as:

$$\dot{H} = c_p T(\dot{m}_{in} - \dot{m}_{out}) \tag{3.10}$$

where T is the absolute temperature of the gas. Naturally the work done calculates as:

$$\dot{W} = p\dot{V} \tag{3.11}$$

The prior math has only performed for one cylinder chamber since the procedure is generic up to this point. Now substituting Equations (3.8), (3.10) and (3.11) in (3.5) and introducing $\kappa = c_p/c_v$ as the adiabatic constant for air and R as the universal gas constant, the pressure in either cylinder chamber can be expressed as:

$$\dot{p}_1 = \frac{\kappa}{V_1}\left(RT\dot{m}_1 - p_1\dot{V}_1\right) \tag{3.12}$$

$$\dot{p}_2 = \frac{\kappa}{V_2}\left(RT\dot{m}_2 - p_2\dot{V}_2\right) \tag{3.13}$$

The coordinate system is now introduced according to Figure 3.1. The volumes of the two chambers are:

$$V_1 = A_1 x_1 \quad V_2 = A_2\left(l - x_1\right) \tag{3.14}$$

the rate of change of the volumes due to the movement of the piston becomes:

$$\dot{V}_1 = A_1\dot{x}_1 \quad \dot{V}_2 = -A_2\dot{x}_1 \tag{3.15}$$

Accordingly one receives the final formulation for chamber A (chamber 1):

$$\dot{p}_1 = \frac{\kappa}{V_1}\left(RT\dot{m}_1 - p_1 A_1\dot{x}_1\right) \tag{3.16}$$

and respectively for chamber B (chamber 2):

$$\dot{p}_2 = \frac{\kappa}{V_2} \left(RT\dot{m}_2 + p_2 A_2 \dot{x}_1 \right) \tag{3.17}$$

which concludes the modeling section for the differential pneumatic cylinder. Subject to further discussion in the next section shall be the modeling of the magnetic valves and the construction of $\dot{m}_1$ and $\dot{m}_2$.

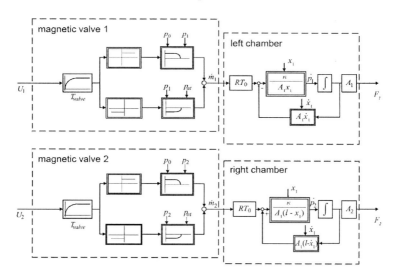

Figure 3.2: Block diagram of pneumatic subsystem [40]

3.2.3 Digitally controlled valves

The magnetic valve is a crucial component of every actuator system since it is the commanding element. It is commonly modeled as an isentropic process leading to an mass flow rate expression for an converging nozzle [78]. Barth et al. [9] introduced digital switching valves with an averaging procedure to cope with the model discontinuities. Richer and Hurmuzlu [65] present a rather detailed model for an proportional spool valve as part of a high performance pneumatic force actuator system.

The convenience of fast switching digitally controlled valves that are discussed in this thesis is that the valve orifice A_v is independent of operating conditions and therefore only takes values of 0 or 1. This assumption holds since the valves have very high dynamics which – due to constructive measures – is not influenced by the acting pressure on each side. Furthermore the valve feeding pressure is constant, being either the air supply pressure or the atmospheric pressure. Resulting in a quasi-linear flow characteristic this simplifies modeling significantly.

Figure 3.3: A module formed of two digitally controlled valves

Since the theory of pneumatic valves is quite established, just the result for the mass flow equation shall be presented according to [65]:

$$
\dot{m}_v = \begin{cases} C_f A_v C_1 \dfrac{p_u}{\sqrt{T}} & \text{if} \quad \dfrac{p_d}{p_u} \leq P_{cr} \\[3mm] C_f A_v C_2 \dfrac{p_u}{\sqrt{T}} \left(\dfrac{p_d}{p_u}\right)^{1/\kappa} \sqrt{1 - \left(\dfrac{p_d}{p_u}\right)^{(\kappa-1)/\kappa}} & \text{if} \quad \dfrac{p_d}{p_u} > P_{cr} \end{cases} \tag{3.18}
$$

where A_v is the valve orifice, $\dot{m}_v$ the mass flow through the orifice, P_u and P_d the upstream and downstream pressures respectively. C_f is the non-dimensional discharge coefficient and C_1, C_2 constants that simplify the notation and calculate as:

$$
C_1 = \sqrt{\frac{\kappa}{R}\left(\frac{2}{\kappa+1}\right)^{\frac{\kappa+1}{\kappa-1}}} \qquad C_2 = \sqrt{\frac{2\kappa}{R(\kappa-1)}} \tag{3.19}
$$

P_{cr} denotes the quotient that separates undercritical and overcritical flow regimes:

$$
P_{cr} = \left(\frac{2}{\kappa+1}\right)^{\frac{\kappa}{\kappa-1}} \tag{3.20}
$$

For air with an adiabatic constant of $\kappa = 1.4$ the prior constants calculate to:

$$
C_1 = 0.040418 \quad C_2 = 0.156174 \quad P_{cr} = 0.528 \tag{3.21}
$$

For the more intuitive characterization of the valves and as a first step to model simplification the mass flow Equations (3.18) is reformulated employing the flow function Ψ:

$$\Psi = \sqrt{\frac{\kappa}{\kappa - 1}\left[\left(\frac{p_d}{p_u}\right)^{\frac{2}{\kappa}} - \left(\frac{p_d}{p_u}\right)^{\frac{\kappa+1}{\kappa}}\right]} \qquad (3.22)$$

which gives:

$$\dot{m} = A_v \Psi p_u \sqrt{\frac{2}{RT}} \qquad (3.23)$$

The flow function Ψ depicted in Figure 3.4 has a maximum value at critical pressure:

$$\Psi(P_{cr}) = 0.484 \qquad (3.24)$$

For practical applications Equation (3.23) can be simplified using b and c, which

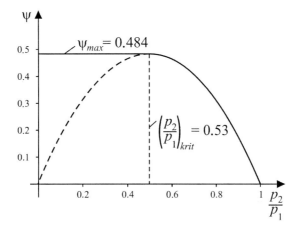

Figure 3.4: Flow function [40]

are determined by experiments [40]:

$$\dot{m}_v = \begin{cases} C_f A_v & if \quad \dfrac{p_d}{p_u} \le b \\[3mm] C_f A_v \sqrt{1 - \left(\dfrac{\frac{p_d}{p_u} - b}{1 - b}\right)^2} & if \quad \dfrac{p_d}{p_u} > b \end{cases} \qquad (3.25)$$

The flow behavior can be visualized as follows: The valve orifice areas A_{v1} and A_{v2} is switched by control inputs u_1 and u_2. The time length from input-activated to valve-open is modeled via a small time delay of $18ms$. A sonic air flow $p_d/p_u \le b$ through the valve orifice is assumed, rendering the mass flow $\dot{m}_v$ to have linear behavior.

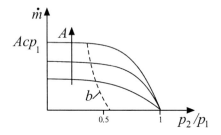

Figure 3.5: Simplified mass flow function [40]

3.2.4 Mechanical subsystem

Figure 3.6 shows a schematic view of the inside of the automated manual transmission of Figure 2.2. The components in the box are the pneumatic cylinder discussed in the previous section and the mechanical components to be discussed as from here. Depicted in Figure 3.7 is the the electro-pneumatic shift actuator's piston rod from

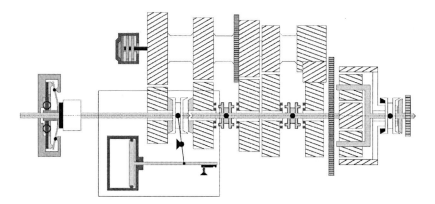

Figure 3.6: Automated manual transmission for heavy commercial vehicles

the actual transmission assembly.

Figure 3.7: Piston rod of the shift actuator

Retaining coil spring

From the sections above it is known that the cylinder chambers are connected to ambient pressure by default, i.e. when the magnetic valves receive no commanding current. Hence the piston position has to be held stable by other means in either of the two end positions. This is done by attaching a coil spring aside the piston exerting a force perpendicularly to the piston's movement. Consider the force F_p to move the piston toward the right as visualized in Figure 3.8. The force F_s exerted

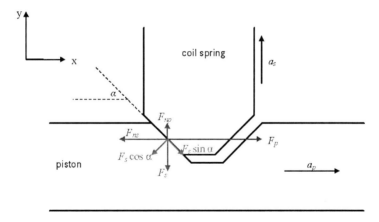

Figure 3.8: Detail of coilspring from Figure 3.6 showing acting forces

on the piston by the coil spring is proportional to the cosine value of the angle α. Thus, the forces which affect the piston are F_p, $F_s \cos \alpha$ and F_{np}, where F_{np} is the normal force on the piston since it can be only moved horizontally. Let F_{pt} be the total force which affects the piston and F_{ptx} and F_{pty} the x- and y-component of

F_{pt} respectively:

$$\vec{F}_{pt} = \vec{F}_p + \vec{F}_s + \vec{F}_{np} \tag{3.26}$$

$$F_{ptx} = F_p - F_s \cos \alpha \sin \alpha \tag{3.27}$$

$$F_{pty} = F_{np} - F_s \cos^2 \alpha = 0 \tag{3.28}$$

The above equations are true as long as the coil spring does not block the movement of the piston i.e. $\alpha_s \neq \alpha_p \tan \alpha$. Otherwise an additional normal force F_{ns} is exerted and slowing down the movement of the piston. Because of the constructive measures taken the normal force F_{ns} can be assumed to be zero.

From the above calculation the characteristic curve for retaining coilspring Figure 3.9 can be constructed. The x-axis represents the force in N and the x-axis the piston position in m. Note that if the coilspring resides in the left notch pushing the piston to the right, the direction of the force is opposite to the situation in which the coil spring resides in the right notch.

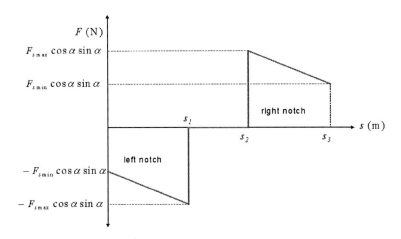

Figure 3.9: Characteristic coil spring curve

Synchronization model

From constructive details of the transmission gears that are operated by the electro-pneumatic shift actuator it is clear that there exist states between the final engaged gears. They were introduced as intermediate positions in Chapter 2. Within these intermediate positions the synchronization of the different rotational speeds of the gear packages takes place, the sprockets of the synchronization inhibiting ring latch. The just described behavior can be seen on the piston position sensor signal since the pneumatic cylinder piston is connected to synchronizer body by a rod. Synchronizing and latching behavior will be described consecutively. Figure 3.10 shows schematic view of a typical synchronizer unit and a shift collar with cogging.

Synchronizer model

The movement of the synchronization body is slowed down by the difference of the rotation speed between the shaft on which the synchronization body is attached and the gear. Before the shaft is locked to the gear, they make frictional contact in order to synchronize their speed. Figure 3.11 shows the synchronization mechanism,

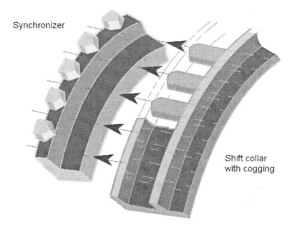

Figure 3.10: Schematic view of a synchronizer [37]

where ω_g is the rotation velocity of the gear, ω_s the rotation velocity of the shaft and F_f the frictional force due to the difference of ω_g and ω_s consequentially. F_p represents the force which acts on the synchronizer-body-piston assembly. F_{pt} is defined as the total force affecting the synchronizer body and F_{ptx}, F_{pty} its components respectively:

$$\vec{F_{pt}} = \vec{F_p} + \vec{F_f} \tag{3.29}$$

$$F_{ptx} = F_f \cos^2 \alpha \tag{3.30}$$

$$F_{pty} = F_p - F_f \cos \alpha \sin \alpha \tag{3.31}$$

The torque M_f is caused by the rotational velocity difference between ω_g and ω_s and the distance from the rotation point to the surface contact r, hence:

$$F_f = \frac{M_f}{r} \tag{3.32}$$

Substituting (3.32) into (3.31) gives:

$$F_{py} = F_p - \frac{M_f \cos \alpha \sin \alpha}{r} \tag{3.33}$$

To be able to calculate the torque M_f from Equation (3.33) the synchronizer is

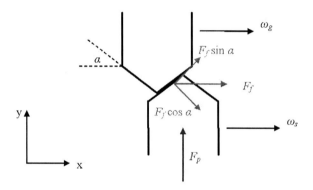

Figure 3.11: Detail of a synchronizer engaging with acting forces

modeled according to Figure 3.12, where C and R are the elasticity and the friction constant respectively: M_{fC} is the torque caused by the elasticity, M_{fR} the torque caused by the friction. They can be calculated as:

$$\dot{M}_{fC} = C\,(\omega_g - \omega_s) \tag{3.34}$$

$$M_{fR} = R\,(\omega_g - \omega_s) \tag{3.35}$$

with (3.34) and (3.35) M_f calculates:

$$M_f = M_{fC} + M_{fR} = C \int (\omega_g - \omega_s)\, dt + R\,(\omega_g - \omega_s) \tag{3.36}$$

Setting M_{fC} as state variable, $\omega_g - \omega_s$ as input, F_f as output and R, C, r as parameters, one gets:

$$\dot{M}_{fC} = C\,(\omega_g - \omega_s)$$
$$F_f = \frac{M_{fC} + R\,(\omega_g - \omega_s)}{r} \tag{3.37}$$

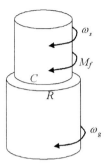

Figure 3.12: Synchronizer rotational model

as a state space system (Figure 3.13).

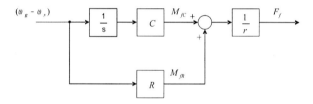

Figure 3.13: Synchronizer block diagram

Inhibiting ring

In the synchronization phase, the inhibiting ring drops in the deepening in the synchronization body. Due to the slope of the deepening, the inhibiting ring slows down the movement of the synchronization body. Figure 3.14 shows the forces which affect the inhibiting ring: It is assumed that the inhibiting ring can only be moved vertically. F_{rt} is the total force which affects the inhibiting ring, F_{rtx} and F_{rty} are the x- and y-component of F_{rt} respectively:

$$\vec{F_{rt}} = \vec{F_r} + \vec{F_p} + \vec{F_{nr}} \tag{3.38}$$

$$F_{rtx} = F_p \sin^2 \alpha - F_{nr} = 0 \tag{3.39}$$

$$F_{rty} = F_p \sin \alpha \cos \alpha - F_r \tag{3.40}$$

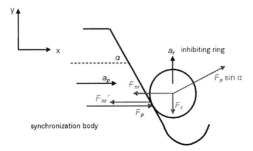

Figure 3.14: Inhibiting ring and acting forces

It can furthermore be assumed that F_r does not affect the synchronization body and the normal force $F'_{nr} = F_{nr}$, i.e. the normal force of the inhibiting ring is transferred completely to the synchronization body. The total force F_{pt} in the synchronization body calculates:

$$F_{pt} = F_p - F'_{nr} = F_p - F_{nr} \tag{3.41}$$

substituting (3.39) in (3.41) gives:

$$F_{pt} = F_p - F_p \sin^2 \alpha = F_p \cos^2 \alpha \tag{3.42}$$

Equation (3.42) is implemented in the model.

Friction components

The above mentioned Coulomb friction F_f splits up into a static friction F_s and a dynamic direction dependent friction F_d [65]:

$$F_f = \begin{cases} F_s & if \quad \dot{x} = 0 \\ F_d \, sign(\dot{x}) & if \quad \dot{x} \neq 0 \end{cases} \tag{3.43}$$

with

$$sign(\dot{x}) = \begin{cases} -1 & if \quad \dot{x} < 0 \\ 0 & if \quad \dot{x} = 0 \\ +1 & if \quad \dot{x} > 0 \end{cases} \tag{3.44}$$

By introducing the assumption $F_s = F_d$ a simplified characteristic friction curve can be constructed.

3.3 Experimental system identification

The main goal of system identification is to derive the structure and parameters of the systems on the basis of observations [84]. In the previous section the system structure was established via theoretical analysis and engineering principles. This section describes the steps undergone to experimentally identify the unknown parameters of the electro-pneumatic shift actuator. From real input and output measurements the parameters of an adjustable model are determined so that they minimize a certain cost function. The adjustment of the model is generally driven by identification algorithms. The combination of both techniques, theoretical analysis and experimental mapping of data, allows a physically meaningful interpretation of the design problem and is often referred as to grey-box-modeling.

3.3.1 Review of identification techniques

System identification surely has been a wide area of research for quite some time now and it is therefore virtually impossible to present in a nutshell. However, some relevant publications shall be noted here. For model estimation problems in a control context the term *System Identification* has originally been formulated by Zadeh [82] in the mid 1950's. Two theoretical main streams evolved from there. The first main stream was the theory of how to realize linear state space models from impulse responses as proposed by Ho and Kalman [34] and followed by Akaike [2], the second one based on prediction errors and time-series analysis as originally outlined by Aström and Bohlin [6]. A theoretically very solid but demanding book was written by Zypkin [84]. Chen and Chang [13] and Ljung [53] give a very recent review of system identification for nonlinear systems in control engineering. Ljung [54] furthermore discusses perspectives on system identification giving insight into the different scientific communities driving the research. Accordingly identification for nonlinear systems is noted to be the most active area today.

3.3.2 Identification grounds

Experimental system identification in this work is performed using Matlab's *System Identification Toolbox, Levenberg-Marquardt Toolbox* and a self designed *brute force algorithm* to gain initial knowledge of the parameter space. The input-output data used for system identification in this thesis was obtained from a test bench that will be described in more detail in Chapter 6. Since the test bench was built using genuine vehicle control modules and since the transmission control module requires lifelike input signals, sinusoidal or pseudo random multi-level signals – which are typical identification input signals – were not permitted to be induced into the test bench. That is why the output data obtained result mainly from step signal inputs on the magnetic valves. A typical set of input-output data that has been used for identification is shown in Figure 3.15. The supply pressure p_v for the pneumatic actuator can be seen from the first subfigure of Figure 3.15. During a shift operation p_v varies only in confines less than 1%. This is due to the large volume of the air supply reservoir compared to the chamber volume of the pneumatic cylinder. The second subfigure of Figure 3.15 shows x_1 being the output of the shift cylinder's position sensor for four inward and outward shifting cycles. As can be seen, x_1 slightly varies from cylce to cycle as there do not exist two completely identical shifting operations. Magnetic valves input signals u_1 and u_2 are depicted in the third subfigure. As stated before the magnetic valves are digital valves driven with a commercial vehicle typical voltage of 24 Volt, wherein 24 Volts translates to logical '1' and 0 Volts to logical '0'. The input regime in the third subfigure is typical for realistic shifting operations. The fourth and fifth subfigure of Figure 3.15 show the chamber pressure x_3 of the first chamber A and x_4 of the second chamber B.

The relationship between valve inputs u_1 and u_2, the chamber pressures x_3 and x_4 and the cylinder position x_1 becomes evident from 3.15. If u_1 is set to '1' a mass flow of compressed air begins to enter chamber A after a short delay. Pressure x_3 then constantly rises but the cylinder's piston still rests in it's initial position. If pressure x_3 exerts a force of about $50N$, the piston begins to move resulting in a pressure drop in x_3. Shortly afterwards u_2 is set to '1' and u_1 to '0' resulting in a mass flow of compressed air entering chamber B. This is to produce a counter pressure x_4 on the piston in order to slow it down for synchronization. The counter pressure x_4 slowing down the piston produces the characteristic plateau in signal x_1. After synchronization u_1 is set to '1' to accelerate the piston to complete the shifting motion. To prevent the piston from hitting the wall of cylinder chamber B too forcefully, u_2 is again set to '1'. If the piston reaches its final position u_1 and u_2 are set to '0' leading in both chambers A and B to deaerate.

3.3.3 Identification of the pneumatic subsystem

The geometry of the cylinder is known exactly: $A_1 = 1.429 \times 10^{-3}\ m^2$, $A_2 = 1.81 \times 10^{-3}\ m^2$, $l = 23.5 \times 10^{-3}\ m$. Further parameters exactly known are the magnetic valves' orifice areas $A_{v1} = A_{v2} = 1.1310 \times 10^{-6}\ m^2$. The discharge coefficients for a mass flow entering a respective chamber were determined to be $c_{fAin} = 0.7033$ and $c_{fBin} = 0.6571$ by identification, the discharge coefficients for a mass flow leaving a respective chamber were identified being $c_{fAout} = 1.096$ and $c_{fBout} = 1.018$. Naturally the identified discharge coefficients are not identical as they not only depend on the magnetic valves' orifice areas but also on the specific shape and placement within the valve.

Figure 3.16 now shows the response of a chamber pressure x_3 to a step input of u_1

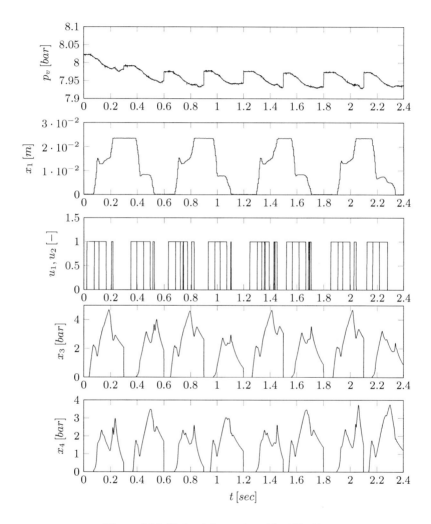

Figure 3.15: Dataset for system identification

for 1.2 sec based on a cylinder model employing the identified discharge coefficients. The supply pressure has been set to 9.1 bar. As can be seen in Figure 3.16 measured and calculated pressures match very well. It can also be seen that the behavior of x_3 in the aerating phases is quite linear up to about 5 bars. In the deaerating phases x_3 exhibits linear behavior down to about 6 bars. In both cases a sonic air flow is present. To validate the identified discharge coefficients an input regime as of

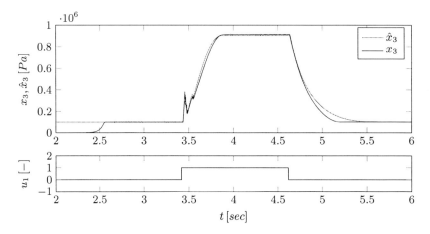

Figure 3.16: Step response of pressure in chamber A

Figure 3.15 was applied. The result can be seen in Figure 3.17 again with a good pressure match.

3.3.4 Identification of the mechanical subsystem

The mass of the piston rod assembly was measured using a digital scale and was found to be $m = 8.598 \ kg$. Known from data-sheets was the coil spring force $F_{ptmax} = 45 \ N$. During identification the Coulomb friction was found to be negligible due to the good lubrication of the piston. The viscous friction coefficient was identified to be $b = 413 \ kgm^{-1}s^{-1}$. As the rotational gear speed ω_g and rotational shaft speed ω_s are synchronized when performing a shifting operation, the maximum counterforce F_f was identified to be $27 \ N$.

Figure 3.19 shows a simulation of the shift cylinder having the identified parameters. Turning to Figure 3.18 it can be seen that the model with the identified parameters resembles the real system quite well.

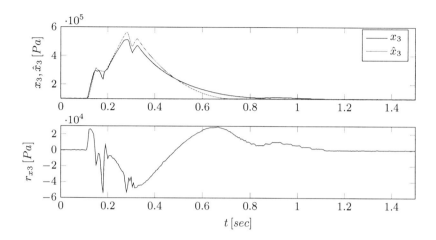

Figure 3.17: Pressure error with identified parameters

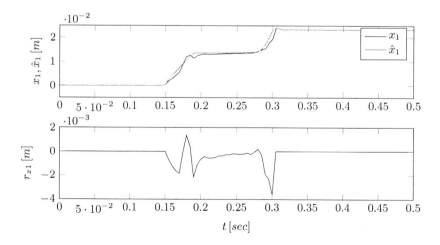

Figure 3.18: Position error with identified parameters

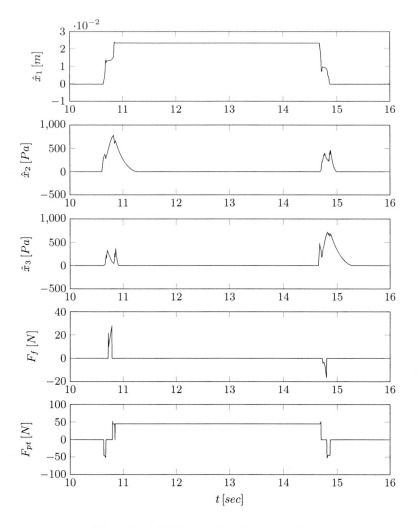

Figure 3.19: Shifting cycle with counter forces

3.3.5 Shift actuator in state space representation

Choosing x_1 as the piston position, x_2 as the piston speed, x_3 as chamber A pressure and x_4 as chamber B pressure, the following state space system including the mass flow rates $\dot{m}_1$ and $\dot{m}_2$ can be constructed:

$$
\begin{aligned}
\dot{x}_1 &= x_2 \\
\dot{x}_2 &= -\frac{F_f}{m} - \frac{F_{pt}}{m} - \frac{b}{m}x_2 + \frac{A_1}{m}x_3 - \frac{A_2}{m}x_4 \\
\dot{x}_3 &= \frac{\kappa}{V_1}\left(RT\dot{m}_1 - x_3A_1x_2\right) \\
\dot{x}_4 &= \frac{\kappa}{V_2}\left(RT\dot{m}_2 + x_4A_2x_2\right)
\end{aligned}
\tag{3.45}
$$

Substituting mass flow rates $\dot{m}_1$ and $\dot{m}_2$ with the above developed valve model one gets the system with inputs u_1 and u_2 accordingly:

$$
\begin{aligned}
\dot{x}_1 &= x_2 \\
\dot{x}_2 &= -\frac{F_f}{m} - \frac{F_{pt}}{m} - \frac{b}{m}x_2 + \frac{A_1}{m}x_3 - \frac{A_2}{m}x_4 \\
\dot{x}_3 &= -\kappa\frac{x_2}{x_1}x_3 - \left(C_fA_\nu\frac{\kappa RT}{A_1x_1}\nu_1\right)u_1 \\
\dot{x}_4 &= \kappa\frac{x_2}{(l-x_1)}x_4 - \left(C_fA_\nu\frac{\kappa RT}{A_2(l-x_1)}\nu_2\right)u_2
\end{aligned}
\tag{3.46}
$$

3.4 Relevant faults and fault construction

A number of faults or faulty behaviors can occur during rugged operation which is common for commercial vehicles. Evidently this is true for the pneumatic shifting components. Nevertheless, from extensive field and database analysis it can be concluded that among a number of faults only a few number of them are likely to occur frequently. A number of faults shall be identified and evaluated due to their practical relevance. On the other hand it has to be noted that not all possible faults can be discussed within this work since they are difficult to be verified in a test bench environment. Table 3.1 lists possible errors in the shifting setup: Especially the leakage induced faults f_1, f_2 and f_3 from the above Table 3.1 have been identified to have high practical relevance in commercial vehicles. Hence, f_1, f_2 and f_3 will be focused on in this thesis.

How the faults f_1, f_2 and f_3 are constructed is depicted in the Figure 6.3 in Chapter 6. An electro-pneumatic shift actuator of the transmission – as can be seen in Figure 6.3 – is equipped with additional connectors leading to bores in the cylinder of the

Component	Fault	Impact
Air supply pipe	low pressure	reduced shifting dynamics $H \leftrightarrow L$
	no pressure	no shifting function
Exhaust pipe	clogged	reduced shifting dynamics $H \leftrightarrow L$
Magnetic valves	stuck	remaining in position H
Shift cylinder	leakage chamber A (f_1)	reduced shifting dynamics $H \rightarrow L$
	leakage chamber B (f_2)	reduced shifting dynamics $L \rightarrow H$
	internal leakage (f_3)	reduced shifting dynamics $L \leftrightarrow H$
	high friction	reduced shifting dynamics $H \leftrightarrow L$

Table 3.1: Shift cylinder related faults

actuator to simulate a pneumatic leakage. The mass flow leaking from either one or between the two chambers of the shift actuator can be measured using a flow-meter and set using attached pneumatic throttles.

4 A fault diagnosis approach based on nonlinear parameter identification

4.1 From system identification to fault diagnosis

System identification and parameter estimation techniques have been applied to fault diagnosis problems as early as in the 70's. Notable research in the 80's was done by Isermann [38] and Patton et al. [62]. A survey paper considering these techniques was presented by Frank [30]. The underlying idea is to generate a residual between model and plant representing model output error or model equation error and applying identification methods to minimize it. This method is based on the assumption that system faults can be linked directly to physical system parameters such as friction, resistance, inductance and so forth. The mathematical model contains an estimated set of parameters, a deviation of these indicates a fault. Initially this concept was applied to static or dynamic models of linear nature but has recently been extended to a larger class of nonlinear systems.

Zhang et al. [83] introduce a system parameter based combined input-output and local approach for fault detection and isolation in nonlinear dynamic systems. They assume that the system under consideration has the same parameterization as its mathematical model. This however is not always the case. Mendoça et al. [59] discuss two different strategies for parameter estimation via a hybrid method. They employ neural network techniques as well as Levenberg-Marquardt algorithm to calculate system parameter deviations. The results are not used for fault diagnosis though. Lee et al. [48] propose a model-based fault detection and isolation method for a robot arm control system where an residual is evaluated using a predefined threshold. Once exceeded, a recursive least square algorithm determines parameter deviations which are handed to a neural network for fault isolation. Huang et al. [35] propose a systematic statistics based approach that employs parameter similarity measures to detect, isolate and identify multiplicative faults in MIMO systems. The similarity measures are based on impulse response sequences and can be implemented for online fault diagnosis. Athamena et al. [7] finally present an fault detection and isolation scheme for an electromagnetic suspension system. Fault detectability and isolability are discussed for the model in discrete-time linear form. Performance considerations when dealing parameter estimation algorithms were presented by Kristensen [47].

4.1.1 Basic diagnosis concept

In this section a fault diagnosis procedure shall be explained as originally presented by Isermann [38]:

1. Choice of a parametric model of the system.

2. Determination of the relationships between the model parameters Θ_i and the physical parameters p_i: $\Theta = f(p)$

3. Identification of the model parameter vector Θ using the input u and output y of the actual system or component

4. Determination of the physical parameter vector $p = f^{-1}\Theta$

5. Calculation of the vector of deviations Δp from its nominal value

6. Decision on a fault by exploiting the relationships between faults and changes in the physical parameters p_i

The diagnosis approach presented above was originally intended for linear models with lumped parameters that can be represented in input/output difference equations [27]. Consider a SISO discrete-time linear system with input $u(k)$ and output $y(k)$:

$$y(k) + a_1 y(k-1) + \ldots + a_n y(k-n) = b_0(u) + b_1 u(k-1) + \ldots + b_m u(k-m) \quad (4.1)$$

with parameter vector Θ and measurement vector $\Phi(k-1)$

$$\begin{aligned}
\Theta &= [a_1 \ldots a_n \; b_1 \ldots b_n]^T \\
\Phi(k-1) &= [y(k-1) \ldots y(k-n) \; u(k-1) \ldots u(k-m)]^T
\end{aligned} \quad (4.2)$$

the system can be rewritten in vector notation as follows:

$$y(k) = \Theta^T \Phi(k-1) \quad (4.3)$$

Since systems faults can be expressed as parameter deviations the system with faults can be rewritten as:

$$y_f(k) = \Theta^T \Phi(k-1) + \Delta\Theta^T \Phi(k-1) = \Theta_f^T \Phi(k-1) \quad (4.4)$$

A cost function $J(k)$ can be defined as follows:

$$J(k) = \sum_{i=1}^{k} (y(k) - y_f(k))^2 \quad (4.5)$$

Minimizing $J(k)$ using *Gauß-Newton* or *Levenberg-Marquardt Algorithm* gives an estimate for Θ_f^T indicating the fault.

The just presented example for a system in linear form is straight forward and can be extended to nonlinear model representations as well. The problem however is to find the exact relationships between the model parameters Θi and the physical parameters p_i. From the numerous experiments that were performed for the electro-pneumatic shift actuator discussed in this work the parametric relationships are known quite well which adds up to the good nature of the problem discussed. Thus, the model parameters Θ_i for the fault induction were chosen to be the flow coefficient analogous to that of the magnetic valves. Therefore, it can be assumed that $\Theta_i \approx p_i$. However, this assumption can not be generalized for arbitrary problems.

4.1.2 Fault detection and isolation hypotheses

Independently from the system nature the fault detection and fault isolation can be defined as follows:

Definition 1. *Fault detection: Let Θ be the real parameter vector of the dynamic system, Θ_0 a vector representing the nominal and Θ_f the values in case of a fault. With a N-sized set of input-output data:*

$$\{(u_k, y_k) \ : \ k = 1, 2, ...N\} \tag{4.6}$$

the decision has to be made between the two hypotheses:

$$\begin{aligned} \mathcal{H}_0 \ &: \ \Theta_0 = \Theta_f \\ \mathcal{H}_1 \ &: \ \Theta_0 \neq \Theta_f \end{aligned} \tag{4.7}$$

Definition 2. *Fault isolation: Let Θ_i be a sub vector of Θ and Θ_0^i as well as Θ_f^i the correspondents suchlike. The fault isolation problem equals the decision between two hypotheses for each Θ_i:*

$$\begin{aligned} \mathcal{H}_0(\Theta_0^i) \ &: \ \Theta_0^i = \Theta_f^i \\ \mathcal{H}_1(\Theta_0^i) \ &: \ \Theta_0^i \neq \Theta_f^i \end{aligned} \tag{4.8}$$

This naturally requires that each fault corresponds to changes in a sub-vector of Θ.

The parameter estimation sequence leading to the decisions to be made above is triggered if a predefined error threshold δ_f is exceeded:

$$F(\boldsymbol{x}) = \frac{1}{2} \sum_{i=1}^{m} (f_i(\boldsymbol{x}))^2 \ > \ \delta_f \tag{4.9}$$

The threshold δ_f will be determined in experiments.

4.1.3 Solving non-linear least square problems

Naturally resulting from the above said, the problem of finding the appropriate fault-related parameter deviations boils down to the problem of minimizing a residual or cost-function. The more difficult part however is to find an appropriate regression model to deal with. This shall be dealt with a little later. Here in the introductory section well established methods for solving least squares problems, namely *Gauß-Newton* and *Levenberg-Marquardt* [49] [56] shall be introduced and briefly discussed. The following is excepted from Madsen et al. [55] who give a very comprehensive insight.

Let $f(x)$ be vector function representing the general residual function with the aim to minimize $\|f(x)\|$, or equivalently:

$$minimize\, \|\boldsymbol{f}(\boldsymbol{x})\| \quad \rightarrow \quad \boldsymbol{x}^* = argmin_x\, \{F(\boldsymbol{x})\}, \tag{4.10}$$

where

$$F(\boldsymbol{x}) = \frac{1}{2}\sum_{i=1}^{m}(f_i(\boldsymbol{x}))^2 = \frac{1}{2}\,\|\boldsymbol{f}(\boldsymbol{x})\|^2 = \frac{1}{2}\boldsymbol{f}(\boldsymbol{x})^T\boldsymbol{f}(\boldsymbol{x}) \tag{4.11}$$

The *Gauss-Newton method* is based on the first derivatives of the components of the vector function i.e. a linear approximation of $\boldsymbol{f}$ in the neighborhood of $\boldsymbol{x}$. For small $\|\boldsymbol{h}\|$ the *Taylor expansion* becomes:

$$\boldsymbol{f}(\boldsymbol{x}+\boldsymbol{h}) \simeq l(\boldsymbol{h}) = \boldsymbol{f}(\boldsymbol{x}) + \boldsymbol{J}(\boldsymbol{x})\boldsymbol{h} \tag{4.12}$$

Inserting (4.12) in (4.11) it is clear that:

$$\begin{aligned} F(\boldsymbol{x}+\boldsymbol{h}) \simeq L(\boldsymbol{h}) &\equiv \frac{1}{2}l(\boldsymbol{h})^T \\ &= \frac{1}{2}\boldsymbol{f}^T\boldsymbol{f} + \frac{1}{2}\boldsymbol{h}^T\boldsymbol{J}^T\boldsymbol{J}\boldsymbol{h} \\ &= F(\boldsymbol{x})\boldsymbol{f} + \frac{1}{2}\boldsymbol{h}^T\boldsymbol{J}^T\boldsymbol{J}\boldsymbol{h} \end{aligned} \tag{4.13}$$

where $\boldsymbol{J}(x)$ is the *Jacobian matrix*. The *Gauß-Newton step* $\boldsymbol{h}_g$ minimizes $L(\boldsymbol{h})$:

$$\boldsymbol{h}_{gn} = argmin_h\, \{L(\boldsymbol{h})\} \tag{4.14}$$

$L(\boldsymbol{h})$ has a minimum which can be found by solving:

$$\left(\boldsymbol{J}^T\boldsymbol{J}\right)\boldsymbol{h}_{gn} = -\boldsymbol{J}^T\boldsymbol{f} \tag{4.15}$$

As an extension to *Gauß-Newton*, *Levenberg-Marquardt* is generally more robust, i.e. has a better convergence for less accurate starting points. Nevertheless, the convergence rate can be slower if the starting point resides close to the minimum. Generally speaking *Levenberg-Marquardt* is a damped method, adding a damping parameter μ to Equation (4.15):

$$\left(J^T J + \mu I\right) h_{lm} = -g, \quad g = J^T f \quad \mu \geq 0 \qquad (4.16)$$

The damping parameter is controlled by the gain ration ρ that is defined as:

$$\rho = \frac{F(x) - F(x + h_{lm})}{L(0) - L(h_{lm})} \qquad (4.17)$$

The the controlling sequence used in this work is:

$$\begin{aligned}
if \; &: \; \rho < 0.25 \\
&\mu := \mu * 2 \\
elseif \; &: \rho > 0.75 \\
&\mu := \mu/3
\end{aligned} \qquad (4.18)$$

This has several desirable effects:

1. For all $\mu > 0$ the coefficient matrix is positive definite, ensuring h_{lm} to be in a descent direction

2. For large μ one gets a short step h_{lm} in the steepest descent direction. This is ideal if the current iterate is far from the solution.

3. If μ is very small i.e. $h_{lm} \cong h_{gn}$ in the final stages of the iteration one gets almost quadratic convergence

The above examples assume that the vector function f is differentiable i.e. the *Jacobian*:

$$J(x) = \left[\frac{\partial f_i}{\partial x_j}\right] \qquad (4.19)$$

exists. In many optimization problems of practical nature – as in the problem at hand – one may not be able to formulate all elements of $J(x)$ exhaustively, because their are either unknown or too complex to be handled analytically. For these cases the *Secant-Levenberg-Marquardt method* can be applied. Here elements of $J(x)$ can be replaced by numerically approximated values:

$$\frac{\partial f_i}{\partial x_j}(x) \simeq \frac{f_1(x + \delta e_j) - f_i(x)}{\delta} \qquad (4.20)$$

where δ is an appropriately small real number.

4.2 Problem formulation

4.2.1 System description and fault induction

Recall above Chapter 3 where the model of the pneumatic shift actuator was presented in the general input affine form:

$$\dot{x} = f(x) + g(x)u$$
$$y = h(x) \tag{4.21}$$

Note that the term $f(x)$ represents the mechanical part of the pneumatic shift actuator, wherein the term $g(x)u$ represents the respective pneumatic part. Since the actuator faults f_1, f_2 and f_3 considered are pneumatic faults strictly, the parameter vector Θ and the faulty parameter vector Θ_f will be only represented in the term the term $g(x)u$. Thus, the faulty system reads as:

$$\dot{x} = f(x) + g(x, \Theta_f)u$$
$$y = h(x) \tag{4.22}$$

where Θ_f expands to:

$$\Theta_f = \begin{bmatrix} \Theta_{f_1} \\ \Theta_{f_2} \\ \Theta_{f_3} \end{bmatrix} = \begin{bmatrix} \omega_1 \cdot (C_f A_v)_{f_1} \\ \omega_2 \cdot (C_f A_v)_{f_2} \\ \omega_3 \cdot (C_f A_v)_{f_3} \end{bmatrix} \tag{4.23}$$

Note that $0 \leq \omega_n \leq 1$ is the fault induction, which will be described in the following with respect to the state space notation derived in Chapter 3. First recall the model of the differential pneumatic cylinder for the respective chamber pressures:

$$\dot{x}_3 = \frac{\kappa}{V_1}(RT\dot{m}_1 - x_3 A_1 x_2)$$
$$\dot{x}_4 = \frac{\kappa}{V_2}(RT\dot{m}_2 + x_4 A_2 x_2) \tag{4.24}$$

and the model with the modeled magnetic valves accordingly:

$$\dot{x}_3 = -\kappa \frac{x_2}{x_1} x_3 - \left(C_f A_v \frac{\kappa RT}{A_1 x_1} \nu_1 \right) u_1$$
$$\dot{x}_4 = \kappa \frac{x_2}{(l - x_1)} x_4 - \left(C_f A_v \frac{\kappa RT}{A_2(l - x_1)} \nu_2 \right) u_2 \tag{4.25}$$

Now consider the third fault f_3, which is characterized by a mass flow $\dot{m}_{L3}$ crossing from chamber A to chamber B or vice versa. Thus $\dot{m}_{L3}$ is introduced to both pressures x_3 and x_4 as can be seen in Equation (4.26):

$$\dot{x}_3 = \frac{\kappa}{V_1} \left(RT \left(\dot{m}_1 - \dot{m}_{L3} \right) - x_3 A_1 x_2 \right)$$
$$\dot{x}_4 = \frac{\kappa}{V_2} \left(RT \left(\dot{m}_2 + \dot{m}_{L3} \right) + x_4 A_2 x_2 \right)$$

(4.26)

Now consider f_1 as the first fault, a wear off of the piston-rod seal resulting in a pneumatic leakage. Naturally this fault only influences the behavior of pressure x_3 in chamber A. In Equation (4.27) it can therefore be represented by a negative mass flow $\dot{m}_{L1}$ leaking out of the chamber. This term can never change sign since the pressure outside the cylinder chamber is always atmospheric pressure p_a and therefore always lower or equal to x_3. As can also be seen from Equation (4.27) fault f_1 has no influence on pressure x_4 in chamber B.

$$\dot{x}_3 = \frac{\kappa}{V_1} \left(RT \left(\dot{m}_1 - \dot{m}_{L1} \right) - x_3 A_1 x_2 \right)$$
$$\dot{x}_4 = \frac{\kappa}{V_2} \left(RT \dot{m}_2 + x_4 A_2 x_2 \right)$$

(4.27)

The second fault f_2 that is denoted in Equation (4.28) is very similar to fault f_1. Fault f_2 represents a leakage in chamber B. Again a negative mass flow $\dot{m}_{L2}$ is leaking out of the chamber only this time strictly influencing pressure x_4, while pressure x_3 remains unaffected. For the direction of mass flow the above said is valid as well:

$$\dot{x}_3 = \frac{\kappa}{V_1} \left(RT \dot{m}_1 - x_3 A_1 x_2 \right)$$
$$\dot{x}_4 = \frac{\kappa}{V_2} \left(RT \left(\dot{m}_2 - \dot{m}_{L2} \right) + x_4 A_2 x_2 \right)$$

(4.28)

Now the Equations (4.26), (4.27) and (4.26) can be transformed using the the following definition of the mass flow, that has already been introduced in Chapter 3:

$$\dot{m}_v = \begin{cases} C_f A_v C_1 \dfrac{P_u}{\sqrt{T}} & if \quad \dfrac{P_d}{P_u} \leq P_{cr} \\ C_f A_v C_2 \dfrac{P_u}{\sqrt{T}} \left(\dfrac{P_d}{P_u} \right)^{1/\kappa} \sqrt{1 - \left(\dfrac{P_d}{P_u} \right)^{(\kappa-1)/\kappa}} & if \quad \dfrac{P_d}{P_u} > P_{cr} \end{cases}$$

(4.29)

Equation (4.26) with induction for fault f_3 now becomes:

$$\dot{x}_3 = -\kappa \frac{x_2}{x_1} x_3 - \left[\frac{\kappa RT}{A_1 x_1} \left(\underbrace{C_f A_\nu \nu_1}_{\Theta_n} - \omega_3 \cdot \underbrace{C_{f3} A_{f3} \nu_{f3}}_{\Theta_{f3}} \right) \right] u_1$$
$$\dot{x}_4 = \kappa \frac{x_2}{(l - x_1)} x_4 - \left[\frac{\kappa RT}{A_2 (l - x_1)} \left(\underbrace{C_f A_\nu \nu_1}_{\Theta_n} + \omega_3 \cdot \underbrace{C_{f3} A_{f3} \nu_{f3}}_{\Theta_{f3}} \right) \right] u_2$$

(4.30)

Note that the nominal system parameter Θ_n and the parameter for fault f_1 Θ_{f_3} are connected additively, where ω_1 is the induction parameter varied by the Levenberg Marquardt algorithm. Equation (4.27) and (4.28) can be transformed accordingly. Since faults f_1 and f_2 influence only one of the chamber pressures there is only one injection equation respectively. Equation (4.27) transforms to:

$$\dot{x}_3 = -\kappa \frac{x_2}{x_1} x_3 - \left[\frac{\kappa RT}{A_1 x_1} \left(C_f A_\nu \nu_1 - \omega_1 \cdot C_{f1} A_{f1} \nu_{f1} \right) \right] u_1 \tag{4.31}$$

while Equation (4.28) transforms to:

$$\dot{x}_4 = \kappa \frac{x_2}{(l - x_1)} x_4 - \left[\frac{\kappa RT}{A_2 (l - x_1)} \left(C_f A_\nu \nu_1 - \omega_2 \cdot C_{f2} A_{f2} \nu_{f2} \right) \right] u_2 \tag{4.32}$$

with the fault induction parameters ω_1 and ω_2 respectively.

4.2.2 Decision making logic and parameter plausibility

Initially the error threshold δ_f for triggering the identification algorithm is set to 20. If the initial residual r_n produced is below 20, the DML will output the result 'no fault':

$$r_n < 20 \rightarrow no\ fault \tag{4.33}$$

Once the error threshold δ_f is exceeded, the identification algorithm is initiated. The identification algorithm will perform calculations for one fault at a time beginning with fault f_1, followed by fault f_2 which is followed by fault f_3. For the faults f_1, f_2 f_3 a respective Θ_{f1}, Θ_{f2} and Θ_{f3} is calculated. To ensure physical plausibility it is then checked whether or not the calculated Θ_{f1}, Θ_{f2} and Θ_{f3} lie in a predefined range. It is furthermore tested whether the model employing a respective calculated Θ_{f1}, Θ_{f2} and Θ_{f3} halves the residual r_n in value. Then the specific fault is determined using the following logic:

$$\begin{aligned} \Theta_{f1} > \Theta_{f2} \wedge \Theta_{f1} > \Theta_{f3} &\rightarrow fault\ 1 \\ \Theta_{f2} > \Theta_{f1} \wedge \Theta_{f2} > \Theta_{f3} &\rightarrow fault\ 2 \\ \Theta_{f3} > \Theta_{f1} \wedge \Theta_{f3} > \Theta_{f1} &\rightarrow fault\ 3 \end{aligned} \tag{4.34}$$

4.2.3 Algorithmic convergence

Naturally, the Levenberg-Marquardt algorithm has to be provided with good starting parameters in order to converge to the correct minimum. Hence, the parameters gained in the above Chapter 3 will be used as starting parameters for fault diagnosis. On the other hand, the trade-off between parameter precision and computing power is evident. The computing power required depends – amongst other things

– on the number of iterations the algorithm has to perform. Keeping in mind that the diagnosis concept introduced has to be implemented under practical conditions, convergence tests have to be performed in order to provide both acceptable parameter precision and acceptable computing power. Figure 4.1 shows the result of

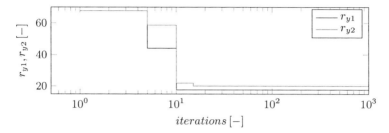

Figure 4.1: Convergence test of Levenberg-Marquardt algorithm

one of the convergence test that were performed. r_{y1} and r_{y2} denote a respective residual sum for two different faults. Table 4.1 is to present the results from Figure 4.1 in more detail. As can be seen, residuals 1 and 2 practically remain constant

iterations	residual y1	residual y2
1	67.827759399868214	67.827759399868214
5	44.117478700272507	58.670064216998234
10	17.646351104396356	21.861282756042723
15	17.646350985709788	20.034212378727968
20	17.646350985709788	20.034212378727968
50	17.646350985709756	20.032727314689161
100	17.646350985709756	20.030805724270081
250	17.646350985709756	20.030647823160088
1000	17.646350985709756	20.030647807335640

Table 4.1: Residual generated in convergence test

after 15 iterations of the Levenberg-Marquardt algorithm. Therefore, the number of iterations for further experiments has been fixed to 20.

4.3 Initial simulation

In the following Figure 4.2 a simulation example for the above introduced diagnosis concept is given. To be able to depict an overall diagnosis cycle in one diagram, the number of iterations for each of the three faults f_1, f_2 and f_3 has been reduced to 8. Consequently the Levenberg-Marquardt algorithm will perform an overall of 24 iterations. The first subplot of Figure 4.2 shows y_1 as the measured piston position and $\hat{y}_1$ as the calculated piston position. A shifting operation of the pneumatic cylinder is iterated 25 times, wherein the first iteration is performed with the Levenberg-Marquardt algorithm deactivated in order to calculate a nominal residual. For the following 24 iterations the Levenberg-Marquardt algorithm is active. The second subplot of Figure 4.2 shows the *cnt* value (solid line), which is a value indicative of which of the three faults f_1, f_2 and f_3 is currently under analysis. Consequently *cnt* is either 1, 2 or 3 in that order. The second subplot of Figure 4.2 also shows the value *imp* (dotted peaks), which is a Boolean indicative of whether or not improvement in the residual was achieved in this very iteration step. The third subplot of Figure 4.2 shows the residuals r_x, being the difference between measured piston y_1 and calculated piston position $\hat{y}_1$, integrated over time. The first step results from the first iteration with the Levenberg-Marquardt algorithm deactivated and is about 56 in the present case. Beginning with the fourth iteration of the cycle designated to fault f_1 the residual r_{f1} for the first fault f_1 begins to decrease notably, coming to rest at about 20. It can also be seen from third subplot of Figure 4.2 that with the beginning with the fourth iteration of the cycle designated to fault f_2 the residual r_{f2} for the second fault f_2 begins to decrease, coming to rest at about 22. Note that the residual r_{f1} for the first fault f_1 remains as the smallest residual. The fourth subplot of Figure 4.2 shows the output of the decision making logic. As the residual for the first fault f_1 was the smallest residual for all iterations, the decision making logic correctly outputs value 1, indicative of that fault f_1 is present in the system.

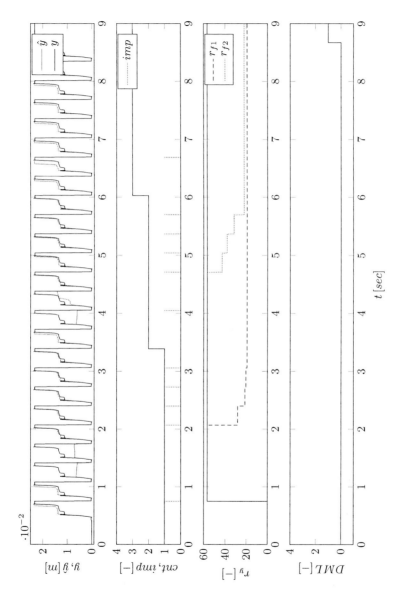

Figure 4.2: Initial simulation of identification based fault diagnosis

5 A fault diagnosis approach based on SMO

5.1 Review of observer tools for nonlinear dynamic systems

The general necessity of observer design originates from the need to reconstruct internal and non-measurable system information from input and output signals for the use in control or diagnosis applications. The concept of state observation for nonlinear systems was originally established by Thau [73] and Kou et al. [44] in the mid 70's and further developed in the following decade by Krener and Isidori [45], Bestle and Zeitz [11], Li and Tao [50] and Walcott and Zak [77]. Most recent research comes from Krishnaswami et al. [46], Barbot et al. [8], Alcorta García and Frank [3], Veluvolo and Soh [74], Veluvolu et al. [75][76] and Floquet and Barbot [29]. The observer designed in this work is mainly inspired by Pandian et al. [61] who applied it in a control context strictly.

Despite the wide and extensive research the problem of nonlinear observation still remains unsolved in terms of a compact theory, which is mainly due to the variety of classes and structures of these nonlinear systems. This underlines the necessity to closely scrutinize the nonlinearities of the model to derive the appropriate observer design procedure.Chen and Saif [20] investigate a general actuator fault isolation problem for linear systems using a bank of observers with corresponding residuals for all possible faulty models, which are of constant nature. The actuator faults can be isolated if only one residual becomes zero, while the others remain nonzero. Chen and Saif [21][22] further demonstrate the observer synthesis using LMIs. Liu et al. [52] propose a fault detection, isolation and estimation scheme for a class of linear systems with external disturbances. An actuator fault is isolated by relating each possible fault to a distinct faulty model, each having attached a sliding mode observer converging to a sufficiently small neighborhood. The results are applied to a civil aircraft model.

Amongst the first to introduce sliding mode observers to diagnosis are Krishnaswami et al. [46]. They establish a model-based approach to the problem of automobile power train monitoring. A discrete sliding mode observer is designed for state and input estimation which are used for a nonlinear parity equation based residual scheme. The approach proves to be effective in both isolating actuator and sensor faults. Yan and Edwards [80] propose a actuator fault detection and isolation scheme for a class of nonlinear systems with uncertainty, which is allowed to have a nonlinear bound being a general function of the state variables. A sliding mode observer is established on the basis of a constrained Lyapunov equation, the equivalent output error injection signal employed to reconstruct the fault signal. Recently a similar approach has been addressed by Wu [79] in the context of a satellite control system.

The first step in the following sections will be to take a closer look at observers for nonlinear systems and their general classification. As an example a simple sliding mode observer design is shown. Secondly, some measures concerning nonlinear observability are introduced concluding this introductory section.

5.1.1 Sliding mode observers

The theory of control of variable structure systems has been developed since the mid 70's. Sliding mode theory for nonlinear systems however has not been introduced until the early 90's by Drakunov [25]. The use of sliding modes for state observation inhibits a number of advantages such as high robustness concerning modeling errors and parameter variations.

A straight forward design procedure for a sliding mode observer will be presented as an introductory example. For convenience the system shall be input affine with a polynomial nature, $g(x, t)$ being globally Lipschitz in x for all $u \in U$, the linear pair (A, c^T) detectable. The systems reads as [41]:

$$\dot{x} = Ax(t) + g(x, t)u$$
$$y(t) = c^T x(t) \tag{5.1}$$

The structure of the observer shall have the following structure with k being a constant gain and $\hat{x}(0) = \hat{x}_0$:

$$\dot{\hat{x}}(t) = A\hat{x}(t) + s(\hat{x}, y, \sigma) + k\left[y(t) - c^T\hat{x}(t)\right] \tag{5.2}$$

Subtracting Equation (5.2) from (5.1) leads to the error dynamics:

$$\dot{\tilde{x}}(t) = \left[A - kc^T\right]\tilde{x}(t) + g(x, u) - s(\hat{x}, y, \sigma) \tag{5.3}$$

with a bounded $s(\hat{x}, y, \sigma)$ and a choice of k such that $A - kc^T$ is asymptotically stable. Using the direct method of Lyapunov, the next step is to find a symmetric, positive definite matrix P fulfilling the matrix Riccati equation

$$F^T P + PF + 2Q = 0 \tag{5.4}$$

and therefore guaranteeing that the Lyapunov function candidate becomes

$$\left.\begin{aligned}\dot{V}(\tilde{x}) &= \frac{1}{2}\tilde{x}^T\left(F^T P + P F\right)\tilde{x}\\&= \tilde{x}^T Q^T \tilde{x}\end{aligned}\right\} \quad > 0 \qquad (5.5)$$

With the calculated k and P now the term $s(\hat{x}, y, \sigma)$ has to be calculated to fulfill the sufficient convergence condition

$$\dot{V}(\tilde{x}) = -\tilde{x}^T Q^T \tilde{x} + \tilde{x}^T P\left[g(x, u) - s(\hat{x}, y, \sigma)\right] < 0 \qquad (5.6)$$

This is guaranteed by the condition of Walcott and Zak [77]

$$\begin{aligned}s(\hat{x}, y, \sigma) &= \begin{cases}\dfrac{P^{-1}cc^T e}{|c^T e|}\sigma(u) & for \quad c^T e \neq 0\\[2mm] 0 & for \quad c^T e = 0\end{cases}\\[2mm]&= P^{-1}c\sigma(u)sign(y - c^T\hat{x})\end{aligned} \qquad (5.7)$$

The scalar function $\sigma(u)$ has to be chosen to fulfill

$$\begin{aligned}\sigma(u) &\geq |\beta(x, u)|\\ g(x, u) &= P^{-1}c\beta(x, u)\end{aligned} \qquad (5.8)$$

The chattering phenomenon induced by $s(\hat{x}, y, \sigma)$ can be reduced by introducing a preferably rather small boundary layer $\epsilon > 0$, the variable observer term therefore calculates as:

$$s(\hat{x}, y, \sigma) = \begin{cases}\dfrac{P^{-1}c\tilde{y}}{|\tilde{y}|}\sigma(\hat{x}, u) & for \quad |\tilde{y}| > \epsilon\\[3mm]\dfrac{P^{-1}c\tilde{y}}{|\epsilon|}\sigma(\hat{x}, u) & for \quad |\tilde{y}| \leq \epsilon\end{cases} \qquad (5.9)$$

5.1.2 Observability measures for nonlinear dynamic systems

General definition of observability

In contradiction to linear systems, a compact theory of designing observers for non-linear systems does not exist, as an observer structure closely depends on system structure. Therefore known observability measures from linear systems can not be applied. Hence, some very general observability measures for nonlinear system, which can also be found in Adamy [1] and Jelali [41], shall be introduced here-inafter.

Definition 3. *Global observability: A system*

$$\dot{x} = f(x, u)$$
$$y = g(x, u)$$

(5.10)

with $x(t_0) = x_0$ and $y \in \mathbb{R}^r$ shall be defined for $x \in D_x \subseteq \mathbb{R}^n$ and $u \in C_u \subseteq C^{n-1}$. The system is called globally observable, if all $x_0 \in D_x$ are determinable from the knowledge of $u(t)$ and $y(t)$ defined over an time interval $[t_0, t_1 < \infty]$ for all $u \in C_u$. Where $\subseteq C^{n-1}$ is the space of the $(n-1)$ times continuously differentiable vector function $u(t)$.

As a similar though weaker condition the local observability shall be defined as:

Definition 4. *Local observability: A system*

$$\dot{x} = f(x, u)$$
$$y = g(x, u)$$

(5.11)

with $x(t_0) = x_0$ and $y \in \mathbb{R}^r$ shall be defined for $x \in D_x \subseteq \mathbb{R}^n$ and $u \in C_u$. The system is called locally observable, if all $x_0 \in D_x$ in an environment

$$U = \{x_0 \in \mathbb{R}^n \ \|x_0 - x_p\| < \rho\}$$

(5.12)

of any $x_p \in D_x$ are determinable from the knowledge of $u(t)$ and $y(t)$ defined over an time interval $[t_0, t_1 < \infty]$ for all $u \in C_u$.

Whereas local observability of linear systems implicates global observability and vice-versa, this is not necessarily true for the nonlinear case. In addition to that observability of nonlinear systems may even depend on input vector u.

Observability of general nonlinear systems

The measure shall now be applied to a general nonlinear system of the form:

$$\dot{x} = f(x, u)$$
$$y = g(x, u)$$

(5.13)

with input vector $\boldsymbol{u}$. As in the procedure for the autonomous systems $n-1$ derivatives are calculated:

$$\dot{y} = \frac{\partial g}{\partial \boldsymbol{x}} \boldsymbol{f}(\boldsymbol{x}, \boldsymbol{u}) + \frac{\partial g}{\partial \boldsymbol{u}} \dot{\boldsymbol{u}} = h_1(\boldsymbol{x}, \boldsymbol{u}, \dot{\boldsymbol{u}})$$

$$\ddot{y} = \frac{\partial h_1}{\partial \boldsymbol{x}} \boldsymbol{f}(\boldsymbol{x}, \boldsymbol{u}) + \frac{\partial h_1}{\partial \boldsymbol{u}} \dot{\boldsymbol{u}} + \frac{\partial h_1}{\partial \dot{\boldsymbol{u}}} \ddot{\boldsymbol{u}} = h_2(\boldsymbol{x}, \boldsymbol{u}, \dot{\boldsymbol{u}}, \ddot{\boldsymbol{u}})$$

$$\vdots \tag{5.14}$$

$$y^{(n-1)} = \frac{\partial h_{n-2}}{\partial \boldsymbol{x}} \boldsymbol{f}(\boldsymbol{x}, \boldsymbol{u}) + \sum_{i=1}^{n-1} \frac{\partial h_{n-2}}{\partial \boldsymbol{u}^{(i-1)}} \boldsymbol{u}^i = h_{(n-1)} \left(\boldsymbol{x}, \boldsymbol{u}, \dot{\boldsymbol{u}}, \ldots, \boldsymbol{u}^{(n-1)} \right)$$

Likewise:

$$\boldsymbol{z} = \begin{bmatrix} y \\ \dot{y} \\ \ddot{y} \\ \vdots \\ y^{(n-1)} \end{bmatrix} = \begin{bmatrix} g(\boldsymbol{x}, \boldsymbol{u}) \\ h_1(\boldsymbol{x}, \boldsymbol{u}, \dot{\boldsymbol{u}}) \\ h_2(\boldsymbol{x}, \boldsymbol{u}, \dot{\boldsymbol{u}}, \ddot{\boldsymbol{u}}) \\ \vdots \\ h_{(n-1)} \left(\boldsymbol{x}, \boldsymbol{u}, \dot{\boldsymbol{u}}, \ldots, \boldsymbol{u}^{(n-1)} \right) \end{bmatrix} = \boldsymbol{q} \left(\boldsymbol{x}, \boldsymbol{u}, \dot{\boldsymbol{u}}, \ldots, \boldsymbol{u}^{(n-1)} \right) \tag{5.15}$$

Definition 5. *Global observability of general nonlinear systems: A system*

$$\dot{\boldsymbol{x}} = \boldsymbol{f}(\boldsymbol{x}, \boldsymbol{u})$$
$$y = g(\boldsymbol{x}, \boldsymbol{u}) \tag{5.16}$$

that is defined on $D_x \subseteq \mathbb{R}^r$ and $C_{\boldsymbol{u}} \subseteq C^{(n-1)}$ is globally observable, if the mapping

$$\boldsymbol{z} = \boldsymbol{q} \left(\boldsymbol{x}, \boldsymbol{u}, \dot{\boldsymbol{u}}, \ldots, \boldsymbol{u}^{(n-1)} \right) \tag{5.17}$$

is uniquely solvable for $\boldsymbol{x}$ for all $\boldsymbol{x} \in D_x$ and $\boldsymbol{u} \in C_{\boldsymbol{u}}$.

Observability rank condition

A considerable criterion for testing whether or not a nonlinear system exhibits local weak observability is the observability rank condition introduced by Hermann and Krener [33].

Definition 6. *Local weak observability of a general nonlinear systems: If a system*

$$\dot{x} = f(x, u)$$
$$y = h(x) \tag{5.18}$$

is locally weakly observable, then the observability rank condition is satisfied generi-

cally, that is if any of the observability matrices

$$\boldsymbol{X}_j = \begin{bmatrix} L_f^0 dh_j \\ L_f^1 dh_j \\ L_f^2 dh_j \\ \vdots \\ L_f^{n-1} dh_j \end{bmatrix} \tag{5.19}$$

is of rank n for $1 \leq j \leq m$.

5.1.3 Model-based fault isolation

As soon as the a fault is detected, the next step is to isolate and thus to clearly locate the fault within the system. Whereas a single residual is sufficient for fault detection, fault isolation usually requires a set of residuals. Almost all model-based fault isolation methodologies can be classified into either the concept of a structured residuals set or a directional residuals set [71]. The two concepts shall be discussed briefly.

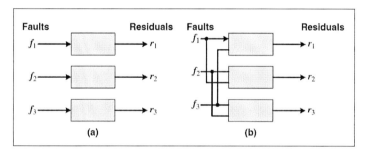

Figure 5.1: Examples of dedicated observer schemes [15]

Structured residual sets

In the structured residuals set based approach, which originates in the ideas of Clark [23], the fault isolation task is to design a set of residuals, where each residual is designed to be sensitive to a subset of faults, while remaining insensitive to the remaining ones. The design procedure can be split up into two steps. The first step is to specify the sensitivity and insensitivity relations between residuals and faults according the requirements of the isolation task. In the second step a set of residual generators addressing these relations have to be constructed. These residual generators again can be constructed using a dedicated or a generalized scheme [15].

An example of how to isolate three faults $[f_1, f_2, f_3]$ in this manner is depicted in Figure 5.1. The three residuals are denoted by $[r_1, r_2, r_3]$. In the dedicated scheme fault isolation scheme, a threshold logic can be used to decide weather a specific fault occurred [15]:

$$r_i(t) > T_i \implies f_i(t) \neq 0; \quad i \in \{1, 2, \ldots, L\} \tag{5.20}$$

where L is the overall number of faults f_i to be isolated and T_i are the thresholds corresponding to the residuals r_i. As can be seen, the dedicated scheme of fault isolation is a straight forward approach, which is rather simple to implement, considering special measures to address uncertainties. If used in a dedicated observer scheme as proposed by [23], an observer uses all system inputs u_i and one chosen output y_k to reconstruct all of the system outputs except the chosen one y_k. By comparing the difference between the estimate and the measurement an indication for a possible fault in the ith sensor is given. If applied to m outputs the technique requires m dedicated observers. Most recently Chen and Saif [14] extended this concept to the isolation of multiple actuator faults, using a bank of observers, with one observer for each actuator under consideration.

Directional residual sets

The usage of directional residual sets addresses the fault isolation task from a geometric point of view. The basic idea is to define a residual space spanned by the residual vector and achieving fault isolation through the design of a directional residual vector, or so called detection filter. A directional residual set is therefore a vector lying in a fixed and a uniquely fault-specific subspace of the overall residual space, the fault itself is determined by the signature direction that is closest to the residual vector generated. The main advantage of the directional residuals set based FDI lies in the rather simple implementation, nevertheless the methodology is strongly problem dependent and requires a number of measures to ensure robustness against modeling errors and system disturbances. The original ideas concerning the fault detection filters came from Beard [10] and Jones [42] and inspired a number of fellow researchers, most notably Massoumnia [57], who first proposed a geometric formulation of the linear type of the Beard-Jones fault detection filter by using the concept of unobservability subspaces. Massoumnia et al. [58] later introduced necessary and sufficient conditions for the FDI problem, the extension to nonlinear systems was carried out by Peris and Isidori [63] proposing a differential-geometric approach. Most recently this approach was extensively applied to a class of electro-hydraulic systems by Chen [17] and Chen and Liu [18][19]. The design methodology can be summarized as followed:

- compute an unobservability distribution implying a state transformation such, that the resulting quotient subsystem is unaffected by all faults except one

- design a fault detection filter for the resulting quotient subsystem

For the nonlinear geometric approach the system is assumed to be in the following form [16]:

$$\dot{x} = f\left(x\right) + \sum_{i=1}^{r} g_i\left(x\right) u_i + \sum_{i=1}^{L} l_i\left(x\right) m_i + \Gamma\left(x\right) w \qquad (5.21)$$

$$y = h(x)$$

where u_i denotes the input channels, m_i the faulty signals to be detected and isolated and w the system disturbances. Unfortunately as proven in a later section this approach can not be applied to the problem in this thesis due to the lack of existence of a state transformation $z = q(x) = T(x)$ or full-state measurement.

5.2 A novel observer scheme for an electro-pneumatic shift actuator

The novel concept introduced hereinafter is based on the idea to design a sliding mode observer that will maintain its sliding motion even if a fault occurs and to come to a fault decision by only evaluating the observer's discontinuity term. However, if in presence of a fault the sliding mode observer maintains its sliding motion, the fault will be compensated by the sliding motion in the same manner as modeling errors and therefore be somewhat masked. By defining thresholds based on the observer's discontinuity term a masked fault can be identified. To the authors best knowledge, such concept has not been presented yet.

To begin with the following design procedure is proposed:

- Perform an observability analysis on state space system

- (if necessary) Perform a system transformation

- Design a main sliding mode observer for the nominal system

- Design auxiliary sliding mode observers for each fault

- Match main observer gain to accommodate auxiliary observers

• Design a fault decision logic based on the main observer's discontinuity term

5.2.1 Observability analysis

The issue of observability of pneumatic actuators has been addressed in very few papers before and, to the authors best knowledge, only in the context of control applications. Wu et al. [78] discuss the possibility of totally eliminating the pressure measurement required for nonlinear control, concluding that only partial observability is achievable. Since the actuator configuration is very similar to the one discussed in this work, aspects of the analysis are transferrable. Girin et al. [32] propose a different design requiring either one of the chamber pressures to be measured, guaranteeing observability at all times. Németh [60] carries out an observability analysis for a pneumatic protection valve as part of an electronic air processing unit used in modern commercial vehicles. Recall the state space representation from Chapter 3 and the observability measure by Hermann and Krener [33] introduced in the previous section. Since neither one of the chamber pressures is measured, the output y is the actuator displacement only, reading

$$y = \begin{bmatrix} 1 & 0 & 0 & 0 \end{bmatrix} \begin{bmatrix} x_1 \\ x_2 \\ x_3 \\ x_4 \end{bmatrix} = x_1 \tag{5.22}$$

$$L_f^0 dh_1 = \frac{\partial h_1}{\partial x} = \begin{bmatrix} 1 & 0 & 0 & 0 \end{bmatrix} \tag{5.23}$$

$$L_f^1 dh_1 = \frac{\partial h_1}{\partial x} \frac{\partial f}{\partial x} + \left[\frac{\partial}{\partial x} \left(\frac{\partial h_1}{\partial x} \right)^T f \right]^T = \begin{bmatrix} 0 & 1 & 0 & 0 \end{bmatrix} \tag{5.24}$$

$$L_f^2 dh_1 = L_f^1 dh_1 \frac{\partial f}{\partial x} + \left[\frac{\partial}{\partial x} \left(L_f^1 dh_1 \right)^T f \right]^T = \begin{bmatrix} 0 & 0 & A_1 & -A_2 \end{bmatrix} \tag{5.25}$$

$$L_f^3 dh_1 = L_f^1 dh_1 \frac{\partial f}{\partial x} + \left[\frac{\partial}{\partial x} \left(L_f^2 dh_1 \right)^T f \right]^T = \begin{bmatrix} M_1 & M_2 & M_3 & M_4 \end{bmatrix} \tag{5.26}$$

The observability matrix $\boldsymbol{X}$ becomes:

$$\boldsymbol{X} = \begin{bmatrix} 1 & 0 & 0 & 0 \\ 0 & 1 & 0 & 0 \\ 0 & 0 & A_1 & -A_2 \\ M_1 & M_2 & M_3 & M_4 \end{bmatrix} \tag{5.27}$$

Hence, the system looses observability if $A_1 M_4 = -A_2 M_3$, which is the case if the cylinder piston is at rest. The whole proof is omitted here due to compactness. Details can be found in [78].

5.2.2 System transformation

Considering the observability analysis above, it can be stated that the system is not uniformly observable if just x_1 is measured. This seems plausible from a physical point of view. System modeling and identification was not the problem since x_3 and x_4 are measurable in the test bench setup. Girin et al. [32] however show that under certain assumption a state transformation is possible if only either one of the chamber pressure is measured. Unfortunately due to the lack of pressure sensors in the pneumatic cylinder discussed in the rather practical application in this work, even this prior assumption of measuring any one of chamber pressure can not be held up. Amongst a number of authors Pandian et al. [61] propose a numerically obtained acceleration to overcome the problem of partial observability. This concept shall be applied in this work as well.

Now recall model of the differential pneumatic cylinder that was derived in Chapter 3:

$$\begin{aligned}
\dot{x}_1 &= x_2 \\
\dot{x}_2 &= -\frac{b}{m}x_2 + \frac{A_1}{m}x_3 - \frac{A_2}{m}x_4 \\
\dot{x}_3 &= \frac{\kappa}{V_1}\left(RT\dot{m}_1 - x_3 A_1 x_2\right) \\
\dot{x}_4 &= \frac{\kappa}{V_2}\left(RT\dot{m}_2 + x_4 A_2 x_2\right)
\end{aligned} \tag{5.28}$$

and the model with the modeled magnetic valves accordingly:

$$\begin{aligned}
\dot{x}_1 &= x_2 \\
\dot{x}_2 &= -\frac{b}{m}x_2 + \frac{A_1}{m}x_3 - \frac{A_2}{m}x_4 \\
\dot{x}_3 &= -\kappa\frac{x_2}{x_1}x_3 - \left(C_f A_\nu \frac{\kappa RT}{A_1 x_1}\nu_1\right)u_1 \\
\dot{x}_4 &= \kappa\frac{x_2}{(l-x_1)}x_4 - \left(C_f A_\nu \frac{\kappa RT}{A_2(l-x_1)}\nu_2\right)u_2
\end{aligned} \tag{5.29}$$

Due to the above mentioned problem of not being able to observe pressures x_3 and x_4 simultaneously, the system order is being reduced to *three* by eliminating control input u_2 and therefore the equation for $\dot{x}_4$, substituting it by an additive term to x_3 that can be interpreted as a system disturbance. One gets an observable system of third order:

$$\dot{x}_1 = x_2$$

$$\dot{x}_2 = -\frac{b}{m}x_2 + \frac{A_1}{m}x_3 + \left[-\frac{A_2}{m}x_4\right]$$

$$\dot{x}_3 = -\kappa\frac{x_2}{x_1}x_3 - \left(C_f A_\nu \frac{\kappa RT}{A_1 x_1}\nu_1\right)u_1$$

(5.30)

$$y = \begin{bmatrix} 1 & 0 & 0 \end{bmatrix}\begin{bmatrix} x_1 \\ x_2 \\ x_3 \end{bmatrix} = x_1$$

(5.31)

Pressure x_4 however remains unmeasurable and unobservable, but shall be determined by numerical methods as proposed above. To describe the dynamics of the differential cylinder in this matter, merely the pressure difference $x_4 - x_3$ is required. From scrutinizing measurements in Chapter 3 rod-assembly mass m and viscous friction coefficient b are known rather precisely. Employing Newton's equation one gets:

$$\dot{x}_2 = -\frac{b}{m} + \frac{A_1}{m}x_3 - \frac{A_2}{m}x_4$$

(5.32)

Since A_1 and A_2 are very close in terms of magnitude $A_3 = \frac{1}{2}(A_1 + A_2)$ shall be introduced enabling one to rewrite Equation (5.32) as:

$$\Delta p = x_3 - x_4 = \frac{m\dot{x}_2 + bx_2}{A_3}$$

(5.33)

which gives the desired chamber pressure difference Δp. To indicate that some numerical uncertainties are involved it shall furthermore be denoted as $\hat{\Delta p}$. Now $\hat{x}_4$ can be obtained as an estimate for the disturbance term in Equation (5.30) by simply subtracting $\hat{\Delta p}$ from the later observed pressure $\hat{x}_3$:

$$\hat{x}_4 = \hat{x}_3 - \hat{\Delta p}$$

(5.34)

An observation error of $\hat{x}_3$ will result in an estimation error of $\hat{x}_4$. This can be compensated by properly tuning the observer for $\hat{x}_3$ or substituting $\hat{x}_4$ by calculating x_4 from u_2. Furthermore it is known from shifting regimes represented by u_1 and u_2 and test bench measurements that x_4 is always atmospheric pressure initially.

5.2.3 Design of a sliding mode observer for leakage detection

To begin with all later used observation errors are defined:

$$
\begin{aligned}
e_1 &= x_1 - \hat{x}_1 \\
e_2 &= x_2 - \hat{x}_2 \\
e_3 &= x_3 - \hat{x}_3
\end{aligned}
\tag{5.35}
$$

Note however that since pressures are not measured, only e_1 and e_2 will account for switching in the following observer, where $[L_1\ L_2\ L_3\ L_4]^T > 0$ are design constants to tune the observer error dynamics and $[K_1\ K_2\ K_3\ K_4]^T$ design constants attached to the sliding mode compensating uncertainties.

$$
\begin{aligned}
\dot{\hat{x}}_1 &= \hat{x}_2 + L_1 e_1 + K_1 sign\,(e_1) \\
\dot{\hat{x}}_2 &= -\frac{b}{m}\hat{x}_2 + \frac{A_1}{m}\hat{x}_3 + \left[-\frac{A_2}{m}\hat{x}_4\right] + L_2 e_2 + K_2 sign\,(e_2) \\
\dot{\hat{x}}_3 &= -\kappa\frac{x_2}{x_1}\hat{x}_3 - \left(C_f A_\nu \frac{\kappa RT}{A_1 x_1}\nu_1\right) u_1 + L_3 e_1 + L_4 e_2 + K_3 sign\,(e_1) + K_4 sign\,(e_2)
\end{aligned}
$$

$$
\tag{5.36}
$$

$$
\begin{aligned}
\dot{e}_1 &= -L_1 e_1 + e_2 - K_1 sign\,(e_1) \\
\dot{e}_2 &= -\left(L_2 + \frac{b}{m}\right)e_2 + \frac{A_1}{m}e_3 - \left[\frac{A_2}{m}e_4\right] - K_2 sign\,(e_2) \\
\dot{e}_3 &= -L_3 e_1 - L_4 e_2 - \kappa\frac{x_2}{x_1}e_3 - K_3 sign\,(e_1) - K_4 sign\,(e_2)
\end{aligned}
\tag{5.37}
$$

Accordingly a simple sliding surface S can be defined with R being the vector of the above defined observation errors and Q a design matrix:

$$
S = QR = \begin{bmatrix} 1 & 0 & 0 \\ 0 & 1 & 0 \end{bmatrix} \begin{bmatrix} e_1 \\ e_2 \\ e_3 \end{bmatrix} = \begin{bmatrix} S_1 \\ S_2 \end{bmatrix}
\tag{5.38}
$$

Naturally the convergence condition for reaching the 2-dimensional sliding surface is:

$$
\begin{aligned}
S_1 \dot{S}_1 &< 0 \\
S_2 \dot{S}_2 &< 0
\end{aligned}
\tag{5.39}
$$

5.2.4 Determination of the observer gains and stability proof

Consequently the just defined notation for the sliding surface should be used for the stability proof. Due to *notational traditions* when employing Lyapunov functions and the straight forward nature of the sliding surface, however, the observation errors e_i are used directly. Consider the first Lyapunov function candidate:

$$V_1 = \frac{1}{2}e_1^2 \tag{5.40}$$

The convergence condition $V_1 \dot{V}_1 < 0$ guarantees that a sliding mode is established. By substitution one receives:

$$\dot{V}_1 = e_1 \left[e_2 - L_1 e_1 - K_1 sign\,(e_1) \right] < 0 \tag{5.41}$$

which translates to:

$$\begin{aligned} K_1 &> e_2 - L_1 e_1 \quad if \quad e_1 > 0 \\ K_1 &> -e_2 + L_1 e_1 \quad if \quad e_1 < 0 \end{aligned} \tag{5.42}$$

as L_1 is strictly positive, the condition $V_1 \dot{V}_1 < 0$ is fulfilled if:

$$K_1 > max\,|e_2| \tag{5.43}$$

The second Lyapunov function is chosen analogously:

$$V_2 = \frac{1}{2}e_2^2 \tag{5.44}$$

which gives:

$$\dot{V}_2 = e_2 \left[-\left(L_2 + \frac{b}{m} \right) e_2 + \frac{A_1}{m} e_3 - \frac{A_2}{m} e_4 - K_2 sign\,(e_2) \right] < 0 \tag{5.45}$$

As can clearly be seen e_3 and e_4 as the pressure observation errors appear in Equation (5.45). To compensate for that it is necessary to tune K_2 significantly high.

Once the above conditions are fulfilled the sliding surface is reached and maintained. Naturally one gets $S_i = 0$ and $\dot{S}_i = 0$. In the achieved sliding mode the stability of the observed pressure can be proven easily applying the *equivalent control method* as introduced by Slotine et al. [70] and later shown by Edwards et al. [26] and Pandian et al. [61]. The main idea is to substitute the discontinuous switching term by a continuous vector M_s. From the definition of the sliding surface (5.38) one gets:

$$\dot{S} = \begin{bmatrix} 1 & 0 & 0 \\ 0 & 1 & 0 \end{bmatrix} \begin{bmatrix} \dot{e}_1 \\ \dot{e}_2 \\ \dot{e}_3 \end{bmatrix} = Q\dot{R} \tag{5.46}$$

The continuous equivalent vector M_s can be generally defined as:

$$M_s = (QK)^{-1} QF \qquad (5.47)$$

where K is a matrix holding the known sliding mode related gains K_i:

$$K = \begin{bmatrix} K_1 & 0 \\ 0 & K_2 \\ K_3 & K_4 \end{bmatrix} \qquad (5.48)$$

and F is a matrix equation connecting the continuous model properties with the pressure observation errors:

$$F = \begin{bmatrix} 0 & 1 & 0 \\ 0 & -\dfrac{b}{m} & \dfrac{A_1}{m} \\ 0 & 0 & -\kappa\dfrac{x_2}{x_1} \end{bmatrix} \begin{bmatrix} 0 \\ 0 \\ e_3 \end{bmatrix} + \begin{bmatrix} 0 \\ -\dfrac{A_2}{m} \\ 0 \end{bmatrix} e_4 \qquad (5.49)$$

After substitution the matrix equation for the equivalent dynamics one the sliding surface becomes:

$$\dot{R} = \left(M_s - K (QK)^{-1} Q \right) F \qquad (5.50)$$

and naturally receive for the pressure observation error:

$$\dot{e}_3 = -\kappa\frac{x_2}{x_1}e_3 - \frac{K_4}{K_2}\frac{A}{m}(e_3 - e_4) \qquad (5.51)$$

which again can be rewritten as:

$$\dot{e}_3 = -\left(\kappa\frac{x_2}{x_1} + \frac{K_4}{K_2} \right) e_3 + \frac{K_4}{K_2}\frac{A}{m}e_4 \qquad (5.52)$$

To complete the stability proof Equation (5.52) has to be discussed thoroughly. The observation error can be stabilized by setting rather high values of K_4/K_2. This also increases the influence of e_4 which on the other hand is bounded and small of magnitude due to the considerations further above. If one takes a closer look at the gains L_i introduced in Equation (5.36) it is clear that large enough L_i result in relatively small observation errors e_i close to the sliding surface. Once the sliding surface is reached the destruction of the sliding mode that could be caused by an increasing e_3 is prevented by the implications of Equation (5.39). It can be concluded that e_3 remains bounded according to Equation (5.52) if as in most cases x_4 is relatively small. In cases of larger x_4 the magnitude of the switching is increased. The quotient x_2/x_1 keeps sign for an unidirectional movement therefore not jeopardizing stability as well.

5.2.5 Fault decision logic based on SMO

The above designed sliding mode observer resembles the nominal behavior of the electro-pneumatic shift cylinder quite well. A deviation from this nominal behavior would be suitable for fault detection, calculating a single residual only. However, since it is desired in this work to distinguish between three different faults, multiple residuals have to be generated.

Naturally, the larger one chooses the gain constant of the switching term, the faster the sliding surface will be approached. However, this may cause undesired chattering of a high frequency around the sliding surface. Another argument for choosing the gain constant of the switching term moderately is to avoid that the fault itself is masked. For simplicity it is assumed that only one of the fault occurs at a time.

Design of auxiliary sliding mode observers

Following the concept of *structured residual sets* introduced above, the first step is to design all possible faulty models. This is straight forward, since modeling has been extensively discussed in Chapter 3. Recall the model equations for $\dot{x}_3$ and $\dot{x}_4$ as the chamber pressures in the mass flow notation first:

$$\begin{aligned}
\dot{x}_1 &= x_2 \\
\dot{x}_2 &= -\frac{b}{m}x_2 + \frac{A_1}{m}x_3 - \frac{A_2}{m}x_4 \\
\dot{x}_3 &= \frac{\kappa}{V_1}\left(RT\dot{m}_1 - x_3 A_1 x_2\right) \\
\dot{x}_4 &= \frac{\kappa}{V_2}\left(RT\dot{m}_2 + x_4 A_2 x_2\right)
\end{aligned} \tag{5.53}$$

In the following the first two of the equations for $\dot{x}_1$ and $\dot{x}_2$ are not redisplayed due to compactness. Now consider f_2 as the second fault, a wear-off of the piston-rod seal resulting in a pneumatic leakage. Naturally this fault only influences the behavior of pressure x_3 in chamber A. In Equation (5.54) it can therefore be represented by a negative mass flow $\dot{m}_{L1}$ leaking out of the chamber. This term can never change sign since the pressure outside the cylinder chamber is always atmospheric pressure p_a and therefore always lower or equal to x_3:

$$\begin{aligned}
\dot{x}_3 &= \frac{\kappa}{V_1}\left(RT\left(\dot{m}_1 - \dot{m}_{L1}\right) - x_3 A_1 x_2\right) \\
\dot{x}_4 &= \frac{\kappa}{V_2}\left(RT\dot{m}_2 + x_4 A_2 x_2\right)
\end{aligned} \tag{5.54}$$

The third fault f_3 that is denoted in Equation (5.55) is very similar to f_2 representing a leakage in chamber B itself. Again a negative mass flow $\dot{m}_{L2}$ is leaking out of the chamber only this time strictly influencing pressure x_4. For the direction of mass flow the above said is valid as well:

$$\dot{x}_3 = \frac{\kappa}{V_1}\left(RT\dot{m}_1 - x_3 A_1 x_2\right)$$
$$\dot{x}_4 = \frac{\kappa}{V_2}\left(RT\left(\dot{m}_2 - \dot{m}_{L2}\right) + x_4 A_2 x_2\right) \tag{5.55}$$

Fault f_1 is characterized by a mass flow $\dot{m}_{L3}$ crossing from chamber A to chamber B or vice versa. Thus $\dot{m}_{L3}$ is introduced to both pressures x_3 and x_4 as can be seen in Equation (5.56):

$$\dot{x}_3 = \frac{\kappa}{V_1}\left(RT\left(\dot{m}_1 - \dot{m}_{L3}\right) - x_3 A_1 x_2\right)$$
$$\dot{x}_4 = \frac{\kappa}{V_2}\left(RT\left(\dot{m}_2 + \dot{m}_{L3}\right) + x_4 A_2 x_2\right) \tag{5.56}$$

Now for each of the just described faulty models an auxiliary sliding mode observer according to Equation (5.36) shall be designed. After performing some transformations one gets the observer for the piston-rod seal leakage:

$$\hat{\dot{x}}_1 = \hat{x}_2 + L_1 e_1 + K_1 sign\left(e_1\right)$$
$$\hat{\dot{x}}_2 = -\frac{b}{m}\hat{x}_2 + \frac{A_1}{m}\hat{x}_3 + \left[-\frac{A_2}{m}\hat{x}_4\right] + L_2 e_2 + K_2 sign\left(e_2\right)$$
$$\hat{\dot{x}}_3 = -\kappa\frac{x_2}{x_1}\hat{x}_3 - \left[\frac{\kappa RT}{A_1 x_1}\left(C_f A_\nu \nu_1 - C_{f1} A_{f1} \nu_{f1}\right)\right]u_1 + L_3 e_1 + L_4 e_2 \tag{5.57}$$
$$+ K_3 sign\left(e_1\right) + K_4 sign\left(e_2\right)$$

the observer for the leakage in cylinder chamber B:

$$\hat{\dot{x}}_1 = \hat{x}_2 + L_1 e_1 + K_1 sign\left(e_1\right)$$
$$\hat{\dot{x}}_2 = -\frac{b}{m}\hat{x}_2 + \frac{A_2}{m}\hat{x}_4 + \left[-\frac{A_1}{m}\hat{x}_3\right] + L_2 e_2 + K_2 sign\left(e_2\right)$$
$$\hat{\dot{x}}_4 = \kappa\frac{x_2}{(l - x_1)}\hat{x}_4 - \left[\frac{\kappa RT}{A_2\left(l - x_1\right)}\left(C_f A_\nu \nu_1 - C_{f2} A_{f2} \nu_{f2}\right)\right]u_2 + L_3 e_1 \tag{5.58}$$
$$+ L_4 e_2 + K_3 sign\left(e_1\right) + K_4 sign\left(e_2\right)$$

and finally the observer for the cross flow leakage:

$$\hat{\dot{x}}_1 = \hat{x}_2 + L_1 e_1 + K_1 sign\,(e_1)$$

$$\hat{\dot{x}}_2 = -\frac{b}{m}\hat{x}_2 + \frac{A_1}{m}\hat{x}_3 + \left[-\frac{A_2}{m}\hat{x}_4\right] + L_2 e_2 + K_2 sign\,(e_2)$$

$$\hat{\dot{x}}_3 = -\kappa\frac{x_2}{x_1}\hat{x}_3 - \left[\frac{\kappa RT}{A_1 x_1}\left(C_f A_\nu \nu_1 - C_{f3} A_{f3} \nu_{f3}\right)\right] u_1 + L_3 e_1 + L_4 e_2 \qquad (5.59)$$

$$\qquad + K_3 sign\,(e_1) + K_4 sign\,(e_2)$$

The stability analysis follows the methodology in Equations (5.38) – (5.52). Naturally ones gets different values for L_i and K_i stabilizing the observers. Note that in observer (5.59) the mass flow $\dot{m}_{L3}$ has an additional influence on x_4 in the opposite chamber. The obtained observer gains L_i and K_i, which are determined by the fault dynamics of f_1, f_2 and f_3 respectively, can be used to tune the sliding mode observer of Equation (5.36) such that it maintains its sliding motion in the presence of a faults. Tuning follows the same procedure as presented in the above Section 5.2.4. Naturally this 'main' observer will perform slightly worse compared to each of the auxiliary.

Fault decision making logic based on SMO discontinuity term

The observer above is designed to maintain its sliding motion even in the presence of any of the faults f_1, f_2 or f_3 respectively. At first to be defined is a value assess the discontinuity term of the sliding mode observer in its sliding phase. Since the observer (5.36) is of third order, three values σ_1, σ_2 and σ_3 have to be defined, wherein:

$$\sigma_1 = \int_{t_0}^{t_{end}} \underbrace{\left[K_1 sign\,(e_1)\right]}_{\psi_1} dt$$

$$\sigma_2 = \int_{t_0}^{t_{end}} \underbrace{\left[K_2 sign\,(e_2)\right]}_{\psi_2} dt \qquad (5.60)$$

$$\sigma_3 = \int_{t_0}^{t_{end}} \underbrace{\left[K_3 sign\,(e_1) + K_4 sign\,(e_2)\right]}_{\psi_3} dt$$

As can be seen σ_1 is the integral of the discontinuity term of the first equation of observer (5.36), σ_2 is the integral of the discontinuity term of the second equation of observer (5.36) and σ_3 is the integral of the discontinuity term of the third equation of observer (5.36), wherein t_0 denotes a respective beginning, t_{end} a respective end of a shifting operation.

The defined values make use of the well known chattering phenomenon occurring in non ideal sliding motions. The reasoning is straight forward. At first a state will be attracted to the sliding surface (5.46) and eventually make contact with it. Since sliding motion is not ideal, the state will not remain but exit the surface on the opposite side, resulting in the switching function to change sign and driving the state back to the surface. Since the respective gains of the sliding mode observer are chosen constant the system being fault free will result in the observed state to chatter about the surface equidistant. Hence,

$$\sigma_1 = \sigma_2 = \sigma_3 = 0 \qquad (5.61)$$

Considering modeling errors it is evident that σ_1, σ_2 and σ_3 will be nonzero. However, it is assumed that the modeling errors are bounded so that in the nominal case

$$|\sigma_1| < \xi_1 \quad |\sigma_2| < \xi_2 \quad |\sigma_3| < \xi_3 \qquad (5.62)$$

wherein ξ_1, ξ_2 and ξ_3 are positive values. An initial values of ξ_1, ξ_2 and ξ_3 can be determined from simulations but should be adjusted in later experiments. Now if any of the faults f_1, f_2 or f_3 occurs in the system, the state will be drawn away from the sliding surface persistently, which again will be counteracted by the discontinuity term accompanied by chattering. However, it has been observed that in this case the state will be further driven away – depending on the magnitude of f_i – from either the positive or the negative side of the sliding surface, resulting in either one of the value σ_1, σ_2 or σ_3 to increase over time, thus qualifying σ_1, σ_2 or σ_3 for fault detection:

$$|\sigma_1| > \xi_1 \vee |\sigma_2| > \xi_2 \vee |\sigma_3| > \xi_3 \rightarrow fault \qquad (5.63)$$

Thus, the fault detection logic of Equation (5.63) is suitable to detect if any of the faults f_1, f_2 or f_3 occurred.

Based on Equations (5.60) and (5.63) a decision making logic can be defined. It is assumed that each σ_1, σ_2 or σ_3 reflects a fault persistently and that only one of the faults f_1, f_2 or f_3 occurs at a time:

$$\{\sigma_1 < -\xi_1\} \wedge \{\sigma_2 > \xi_2\} \wedge \{\sigma_3 < -\xi_3\} \rightarrow fault1$$
$$\{\sigma_1 > \xi_1\} \wedge \{\sigma_2 < -\xi_2\} \wedge \{\sigma_3 < -\xi_3\} \rightarrow fault2 \qquad (5.64)$$
$$\{\sigma_1 < -\xi_1\} \wedge \{\sigma_2 < -\xi_2\} \wedge \{\sigma_3 > \xi_3\} \rightarrow fault3$$

5.3 Simulations

In this section the decision making logic (5.64) introduced above will be simulated for nominal system behavior as well as for each faults f_1, f_2 or f_3. The model parameters are the ones identified in Chapter 3. The observer parameters have been determined in Section 5.2.4.

In each of the respective subsections a simulation with inactive observer will be presented first in order to demonstrate system behavior in the nominal case and of faults f_1, f_2 or f_3 without any influence of the sliding mode observer. Secondly a simulation with active observer will be shown. In each third respective diagram the switching term of the observer for each of faults f_1, f_2 or f_3 is depicted. Each of the respective subsections is concluded with a diagram showing the respective σ_1, σ_2 or σ_3 an their logic conjunction as in (5.64). For the simulation ξ_1, ξ_2 and ξ_3 were preset to '0'.

5.3.1 Experiment 1: Fault free

The first experiment is performed for the fault free case. The results are shown in Figure 5.2 and Figure 5.3. Naturally, since no fault is induced in the modeled system, the observation errors e_1 and e_1 are *zero*, as can be seen in Figure 5.2. Figure 5.3 shows the system behavior with the sliding mode observer being active. Here too, the observation errors e_1 and e_1 remain *zero* at all times.

5.3.2 Experiment 2: Leakage in piston seal (fault 1)

For the second experiment a leakage in the piston seal is induced. The results can be seen in Figures 5.4 to 5.7. First take a look at Figure 5.4, which shows the system behavior with sliding mode observer being inactive. As expected the shifting action is significantly delayed, since the compressed air leaks directly into the opposite pressure chamber. As can be seen from Figure 5.4, position error of the cylinder position rises up to 50%, the pressure error of the cylinder pressure up to 20%. Now the sliding mode observer is set active. The result can be seen in Figure 5.5. The position error of the cylinder position with the sliding mode observer active remains below 2%, the pressure error of the cylinder pressure below 5%. Note that prior to the shifting phase, the pressure error of the cylinder pressure is temporarily about 10%. Figure 5.6 shows the discontinuity term for each state variable x_1, x_2 and x_3. The integral values of the σ_1, σ_2 or σ_3 of each discontinuity term is depicted in Figure 5.7. As can be seen Figure 5.7 σ_1 and σ_3 decrease almost linearly to a value significantly below 0, while σ_2 swings to a value above 0. The decision making logic correctly outputs value 1 indicating that fault f_1 is present in the system. Value 1 is put out after the shifting action is completed.

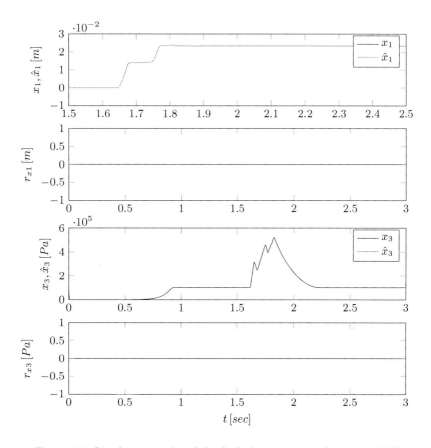

Figure 5.2: Simulation results of the fault free system with inactive SMO

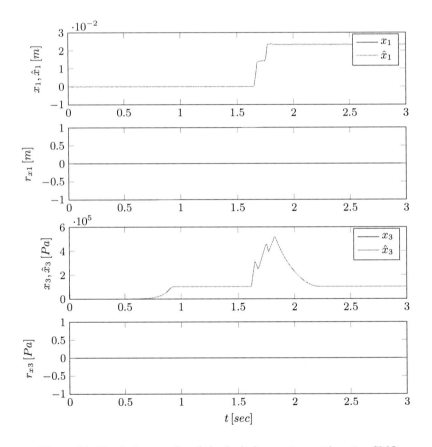

Figure 5.3: Simulation results of the fault free system with active SMO

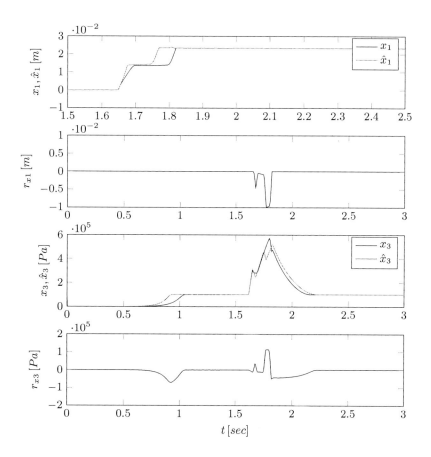

Figure 5.4: Simulation results for fault 1 with inactive SMO

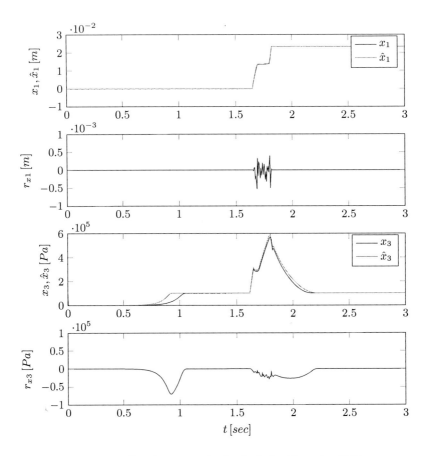

Figure 5.5: Simulation results for fault 1 with active SMO

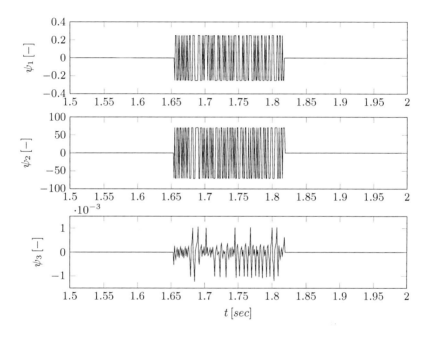

Figure 5.6: Switching of SMO for simulated fault 1

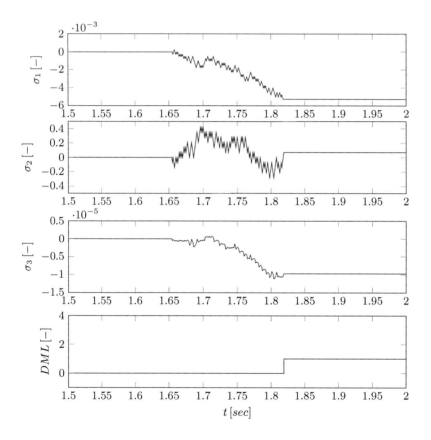

Figure 5.7: Decision making logic for simulated fault 1

5.3.3 Experiment 3: Leakage in piston rod seal (fault 2)

For the third experiment a leakage in the piston rod seal is induced. The results can be seen in Figures 5.8 to 5.11. Figure 5.8 shows the system behavior with sliding mode observer being inactive. As can be seen from 5.4 the shifting action is delayed less then in Figure 5.4 for fault f_1. This is because the compressed air leaks into the surrounding, and not into the opposite pressure chamber. Hence, a much smaller counterforce is presented to the piston. As can be seen from Figure 5.4, position error of the cylinder position rises up to 10%, the pressure error of the cylinder in the shifting phase up to 10%. Now the sliding mode observer is set active. The

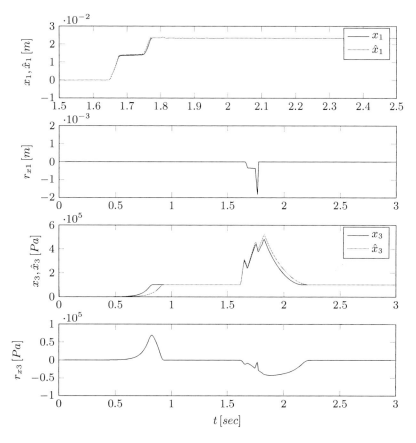

Figure 5.8: Simulation results for fault 2 with inactive SMO

result can be seen in Figure 5.9. The position error of the cylinder position with the sliding mode observer active remains below 2%, the pressure error of the cylinder pressure below 8%. Prior to the shifting phase, the pressure error of the cylinder pressure is temporarily about 12%. Figure 5.10 shows the discontinuity term for

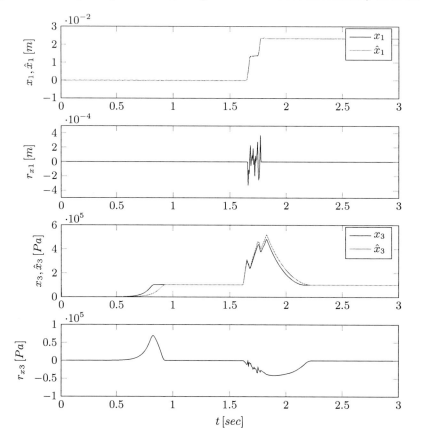

Figure 5.9: Simulation results for fault 2 with active SMO

each state variable x_1, x_2 and x_3. The integral values of the σ_1, σ_2 or σ_3 of each discontinuity term is depicted in Figure 5.11. As can be seen Figure 5.11 σ_2 and σ_3 decrease to a values below 0, while σ_1 swings to a value significantly above 0. After the shifting action is completed, the decision making logic correctly outputs value 2 indicating that fault f_2 is present in the system.

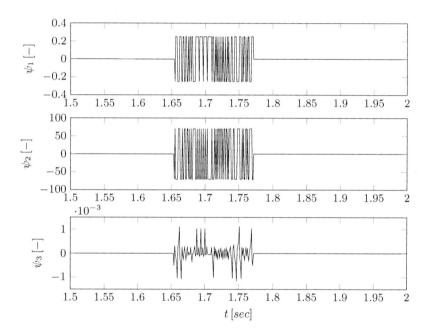

Figure 5.10: Switching of SMO for simulated fault 2

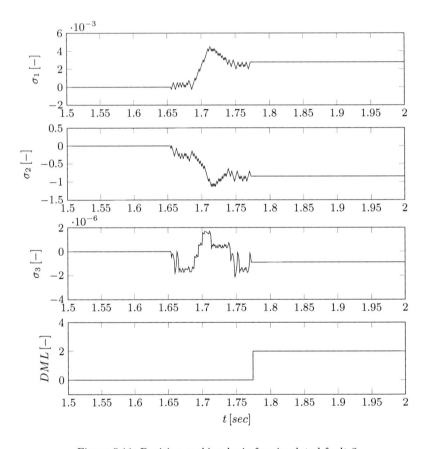

Figure 5.11: Decision making logic for simulated fault 2

5.3.4 Experiment 4: Leakage in cylinder (fault 3)

For the third and last experiment a leakage in the cylinder is induced. The results can be seen in Figures 5.12 to 5.15. Figure 5.12 shows the system behavior with sliding mode observer again being inactive. Not surprisingly, the shifting action is notably faster, since the compressed air leaks from that pressure chamber that counteracts the motion of the cylinder piston. As can be seen from Figure 5.12, the position error of the cylinder position peaks up to 35%. However, contrary to the position error of fault f_1 and f_3, the position error of fault f_1 is positive, indication the faster shifting action. As can also be seen from Figure 5.12, the pressure error of the cylinder pressure up peaks to about 20%. Again the sliding mode observer is set active. The result can be seen in Figure 5.13. The position error of the cylinder position with the sliding mode observer active now remains below 3%, the pressure error of the cylinder pressure below 3%. Figure 5.14 shows the discontinuity term for each state variable x_1, x_2 and x_3. The integral values of the σ_1, σ_2 or σ_3 of each discontinuity term is depicted in Figure 5.15. Looking at Figure 5.11 it can be seen that σ_1 rises almost linearly well above 0, while σ_2 decrease to a value well below 0. σ_1 comes to rest above 0. Therefore, after the shifting action is completed, the decision making logic correctly outputs value 3 indicating that fault f_3 is present in the system.

5.3.5 Conclusion

The simulations presented above confirms the capability of the sliding mode observer to maintain a sliding motion in the presence of each of the induced faults f_1, f_2 or f_3. Furthermore, the designed observer shows good properties for all induced faults f_1, f_2 or f_3. Such, the residual for cylinder position remains below 2%, the residual of the chamber pressure below 5% for all faults. By means of a logical conjunction of σ_1, σ_2 and σ_3, each of the faults f_1, f_2 or f_3 can be detected separately. In Chapter 6 hereinafter, all three fault scenarios will be tested on a realistic test bench environment.

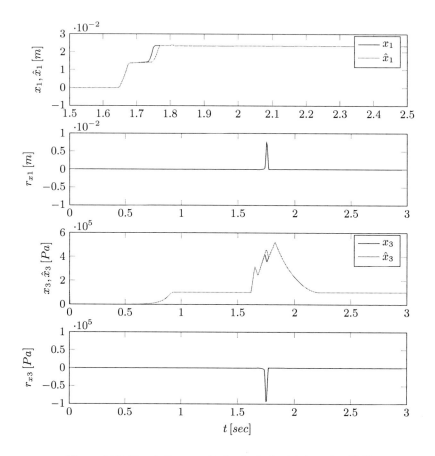

Figure 5.12: Simulation results for fault 3 with inactive SMO

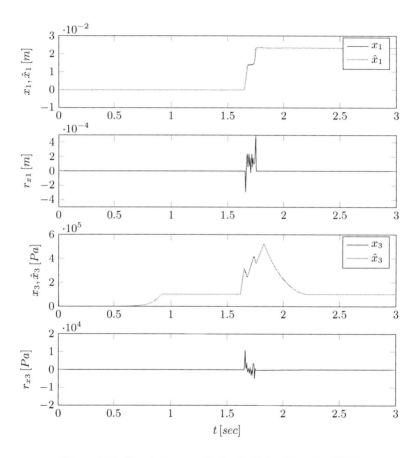

Figure 5.13: Simulation results for fault 3 with active SMO

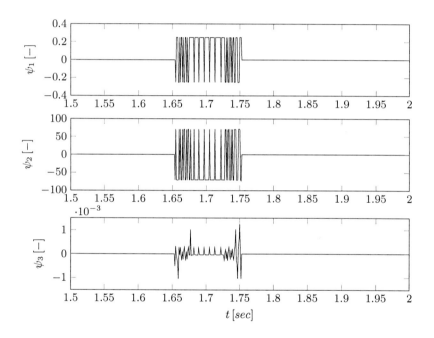

Figure 5.14: Switching of SMO for simulated fault 3

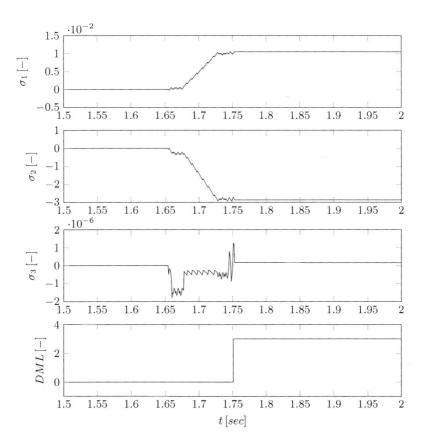

Figure 5.15: Decision making logic for simulated fault 3

6 Practical implementation and experimental results

In this chapter the on-board fault detection and data acquisition unit described
in Section 2.3 and the two off-board diagnosis concepts developed in Chapter 4
and Chapter 5 will be subject to verification. The practical implementation of the
algorithms on concrete hardware will be described first, followed by the experimental
results.

6.1 Practical issues

6.1.1 Hardware and Software

The vehicle on-board fault detection and data acquisition unit was implemented in
MATLAB/Simulink and translated to MISRA-C conform C++ code using MAT-
LAB's Real time workshop. The so generated C++ code was transferred to CarMe-
dialab's mobile development platform 'fleabox' based on Infineon's 32-Bit TriCore
processor. The complete on-board fault detection and data acquisition unit is depicted in Figure 6.1, wherein the 'fleabox' can be seen in the upper left. The box
itself is connected to a combined GPS/GSM antenna (upper left corner of Figure
6.1) by means of which the location of a vehicle carrying said box can be tracked
and measurement data be transferred. On the right hand side of Figure 6.1 a cus-
tomized adapter-cable can be seen via which both the vehicle's diagnosis and power
train buses can be interfaced. The author implemented and installed a respective
on-board unit in six heavy commercial vehicles of the same type for test runs in
Rovaniemi, Finland and Sierra Nevada, Spain. The measurement data obtained
was used to determine shift-duration and fault occurrence statistics and to verify
the usability of the obtained measurement data for fault diagnosis. However, a detailed presentation of the measurement data obtained from vehicle test is beyond
the scope of this thesis.

The experimental results presented in this chapter are based on measurements obtained from a realistic test bench – depicted in Figure 6.2 – that will be described
hereinafter. The test bench also provided the measurements used for identification
in Chapter 3. The platform depicted on the left hand side of Figure 6.2 carries a typical 12-way electronically-pneumatically shifted constant mesh transmission with an
automated dry clutch used for heavy commercial vehicles. The transmission input
is driven by a 11-kW asynchronous motor, while the transmission output can be

dragged using a 3-kW asynchronous motor to simulate a vehicle's rolling friction. Both motors are mounted within the base of the platform for safety reasons. An electro-pneumatic shift actuator of the transmission – as can be seen in Figure 6.3 – has been equipped with additional connectors leading to bores in the cylinder of the actuator to simulate a pneumatic leakage. The mass flow leaking from either one chamber or between the two chambers of the shift actuator can adjusted using attached pneumatic throttles and measured using a flow-meter. The platform depicted on the right hand side of Figure 6.2 carries (from left to right): CANoe based control equipment for controlling the asynchronous motors connected to the transmission and for simulating vehicle bus communication, a typical dashboard of a commercial vehicle with a gear lever for effecting a gear change and for visualizing vehicle speeds, and finally a terminal for operating CANape/IPETRONIK based measurement equipment to be able to measure shift actuator position and chamber pressures and to perform fault diagnosis. To be able to test fault diagnosis under workshop conditions, the diagnosis concept developed in Chapter 4 was exemplarily implemented on a typical diagnostic-tester used in the automotive industry as shown in Figure 6.4. The arrow in the lower midst of Figure 6.4 points at a leakage of an electro-pneumatic shift actuator, such as the leakage of the electro-pneumatic shift actuator depicted in Figure 6.3.

6.1.2 Parameter settings

The sampling time is chosen as $T_s = 1\ ms$. The system parameters for the diagnosis calculations are the ones obtained in identification in Chapter 3. The supply pressure p_v was set to 8.5 bar. The error threshold δ_f for triggering the identification algorithm was set to 20. The lower plausibility bound for Θ_{f1}, Θ_{f2} and Θ_{f3} was chosen to be 0.1858×10^{-6}, the respective upper plausibility bound 0.7432×10^{-5}. The factor for the minimum improvement of the residual was set to 0.5.

The sliding mode observer was tuned as follows:

$$L_1 = 1.9 \quad L_2 = 2.5 \quad L_3 = 0.001 \quad L_4 = 0.01 \tag{6.1}$$

$$K_1 = 0.25 \quad K_2 = 70 \quad K_3 = 0.0005 \quad K_4 = 0.0005 \tag{6.2}$$

The decision making logic of the sliding mode observer uses the following parameters:

$$\xi_1 = 5.00 \times 10^{-3} \quad \xi_2 = 0.35 \quad \xi_3 = 5.00 \times 10^{-6} \tag{6.3}$$

A pneumatic leakage for each the faults f_1, f_2 f_3 was experimentally adjusted to about 25 % using precision throttle valves as can be seen in the lower right corner of the left subfigure of Figure 6.3. The pneumatic leakage was adjusted so as to achieve a noticeable slow down in shift cylinder speed while still allowing the shift cylinder to complete its shifting operation. In order to ensure comparable

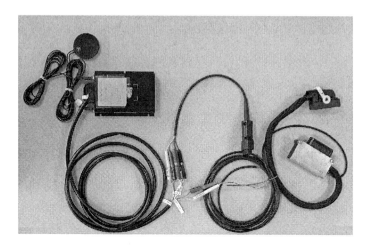

Figure 6.1: On-board fault detection and data acquisition unit

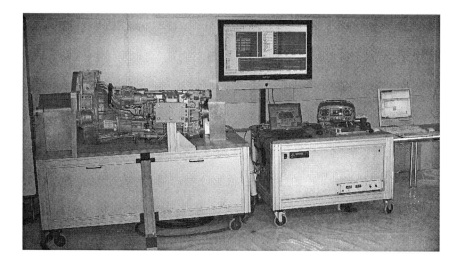

Figure 6.2: Automated manual transmission test bench

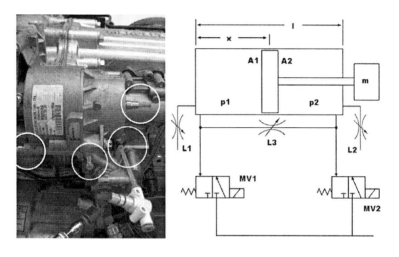

Figure 6.3: Electro-pneumatic shift actuator with simulated leakage

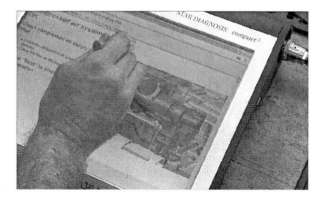

Figure 6.4: Testing unit

measuring results, transmission input and output motors were operated such that
no significant counter-torque originating from the gearbox acted on the shift cylinder
in operation.

6.2 FDI concept based on nonlinear parameter identification

It should be noted that the identification procedure will perform fault calculations
for one fault at a time beginning with fault f_1, followed by fault f_2 which is followed
by fault f_3. Consequently there will be an overall of 61 iterations including the
initial iteration. To ensure 'non-crowded' diagrams only the first 10 iterations for
each fault will be displayed. This seems acceptable since numerous experiments did
show that the Levenberg-Marquardt converges 'steepest' within the first 8 iteration,
while being rather static thereinafter. With regard to the decision making logic
only fault f_1 and fault f_2 were considered for the inward shift cylinder, as fault f_3
is directed at a leakage in cylinder B.

6.2.1 Experiment 1: Fault free

Figure 6.5 shows a typical example of the fault free case. As can bee seen from the
third subfigure of Figure 6.5 the shift duration is 160 ms. It can be seen from the
first subfigure of Figure 6.5 that only one (the first) iteration is present, wherein $\hat{y}$
already fits y very well. The initial residual r_i – the first step in the second subfigure
– is below the residual threshold of 20. Consequently the decision making logic – the
fourth subfigure of Figure 6.5 – display value 0 indicative of that no fault occurred
without initiating the identification algorithm.

6.2.2 Experiment 2: Leakage in piston seal (fault 1)

A typical example of the shift cylinder having a leakage in piston seal is shown in
Figure 6.6. As can bee seen from the third subfigure of Figure 6.6 the shift duration
is 200 ms, which is 125 % slower than the shift duration of the fault free cylinder. In
the first iteration (where the identification is not yet started) there is a rather large
difference between $\hat{y}$ and y in the second shifting phase. The is physically plausible
since in fault f_1 the mass flow from the first chamber leaks into second chamber
resulting in a gradual increase in pneumatic counter pressure. The initial residual
r_i calculates to 67.8. Thus, the Levenberg-Marquardt algorithm is initiated – as
can also be seen from the iteration cycles in the first subfigure.

Turning to that subfigure it can be seen that for the second iteration beginning

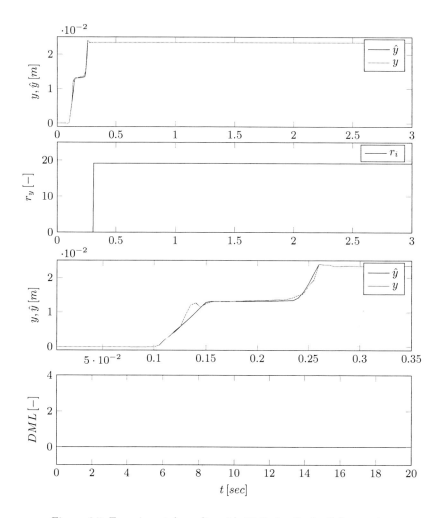

Figure 6.5: Experimental results with DML for the fault free system

at about 0.7 s the fit between $\hat{y}$ and y became significantly worse. This trend slightly improves with the third iteration beginning at about 1.0 s. After the fourth iteration the residual r_{f1} begins to decrease with every iteration. As can be seen in the second subfigure, the residual r_{f1} is below 20 after the ninth iteration. The final residual for iteration cycle for fault f_1 calculates to 17.646, the final residual r_{f2} for the iteration cycle for fault f_2 to 20.034. The fourth subfigure of Figure 6.6 correctly displays value 1 indicative of that fault f_1 occurred. Fault parameter Θ_{f1} was determined to be 3.647×10^{-7}, which is a 32.25 % error to the nominal Θ_{n_1}.

6.2.3 Experiment 3: Leakage in piston rod seal (fault 2)

Figure 6.7 depicts a typical example of the shift cylinder having a leakage in piston rod. As can bee seen from the third subfigure of Figure 6.7 the shift duration is 220 ms, which is 137 % slower than the shift duration of the fault free cylinder. In the first iteration the difference between $\hat{y}$ and y is slightly less then expected but present both the first and the second shifting phase. The is plausible since in fault f_2 the mass flow from the first chamber leaks directly into the surrounding, resulting in no pneumatic counter pressure from chamber B. The initial residual r_i calculates to 53.8. Thus, the Levenberg-Marquardt algorithm is initiated. Turning to that subfigure it can be seen that for the second iteration beginning at about 0.7 s the misfit between $\hat{y}$ and y is quite large. This trend significantly improves with the third iteration beginning at about 1.0 s. After the fourth iteration the residual r_{f2} begins to decrease with every iteration.

As can be seen in the second subfigure, the residual r_{f2} is below 20 after the ninth iteration. Note that in the second subfigure the residual r_{f1} for fault f_1 remains steady at 20.03. This is because Figure 6.7 depicts the cycle for the second fault f_2. Prior to that and not shown in Figure 6.7 the calculations for fault f_1 have already been performed with final residual r_{f1} of 20.03. However, the final residual r_{f2} for the iteration cycle for fault f_2 is 19.25. The fourth subfigure of Figure 6.7 correctly displays value 2 indicative of that fault f_2 occurred. Fault parameter Θ_{f2} was determined to be 4.0309×10^{-7}, which is a 36.64 % error to the nominal Θ_{n_2}.

6.2.4 Experiment 4: Leakage in cylinder (fault 3)

A typical example of the shift cylinder having a leakage in cylinder B is shown in Figure 6.8. As can bee seen from the third subfigure of Figure 6.8 the shift duration is 300 ms, which is 187 % slower than the sift duration of the fault free cylinder. The initial residual r_i calculates to 104.16, resulting in the Levenberg-Marquardt algorithm to be initiated. Turning to that subfigure it can be seen that already

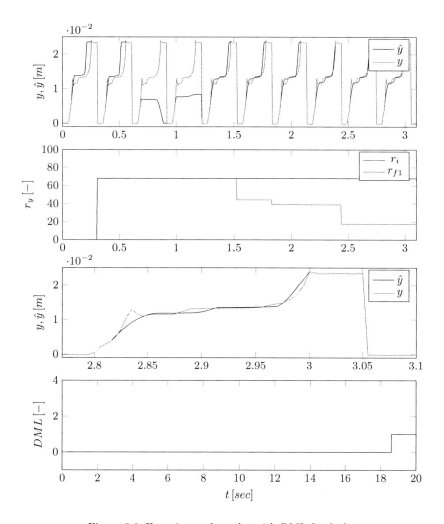

Figure 6.6: Experimental results with DML for fault 1

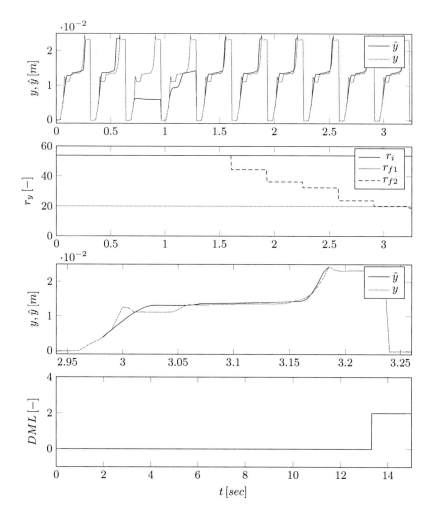

Figure 6.7: Experimental results with DML for fault 2

after the third iteration the fit between $\hat{y}$ and y is very good.

As can be seen in the second subfigure, the final residual r_{f1} of fault f_1 remains steady at 51.79, the final residual r_{f2} of fault f_2 at 89.73. Naturally this is the case as Figure 6.8 depicts the cycle for the third fault f_3. Prior to that and not shown Figure 6.8 the calculations for fault f_1 and f_2 have already been performed. As can be seen the final residual r_{f3} for the iteration cycle for fault f_3 drops down to 25.79. Thus, the fourth subfigure of Figure 6.8 correctly displays value 3 indicative of that fault f_3 occurred. Fault parameter Θ_{f3} was determined to be 1.337×10^{-6}, which is a 118.25 % error to the nominal Θ_{n_3}.

6.3 FDI concept based on sliding mode observer

6.3.1 Experiment 1: Fault free

At first the sliding mode observer shall applied to the fault free case, that is no leakage present in the system. Figures 6.9 and 6.10 show a typical example accordingly. The first subfigure of Figure 6.9 shows good fit between x_1 and $\hat{x}_1$ for nominal inward shifting motion of the actuator. Turning to second subfigure of Figure 6.9, which shows the residual $x_1 - \hat{x}_1$, it can be seen that the residual remains less then $2.7 \times 10^{-3} m$, which corresponds to about 12% with respect to the maximum shift of the actuator. Note that the residual is as expected not smooth but exhibits quite an amount of chattering. Subfigure three shows x_3 and $\hat{x}_3$ being the measured and estimated pressure of the actuator's chamber A. As can be seen from the fourth subfigure, the residual $x_3 - \hat{x}_3$ remains below $1.17 \times 10^{-5} Pa$ in the shifting phase , which corresponds to about 21% with respect to the maximum chamber pressure of chamber the actuator. The effect of the observers chattering can be observed in the pressure residual as well.

Recall that the observer gains were chosen constant so as to allow for the observer to maintain its sliding motion even if a fault occurs. However, this has the effect that the observer performs slightly 'worse' in the fault free case. Figure 6.10 shows in the first three subfigures σ_1, σ_2 and σ_3 being the values characterizing the sliding mode observer's discontinuity term and calculating to:

$$\sigma_1 = -2.69 \times 10^{-3} \quad \sigma_2 = 2.15 \times 10^{-1} \quad \sigma_3 = 1.92 \times 10^{-6} \tag{6.4}$$

The output of the decision making logic is depicted in subfigure four. As no fault is present in the system the decision making logic outputs value 0 indicating the fault free case.

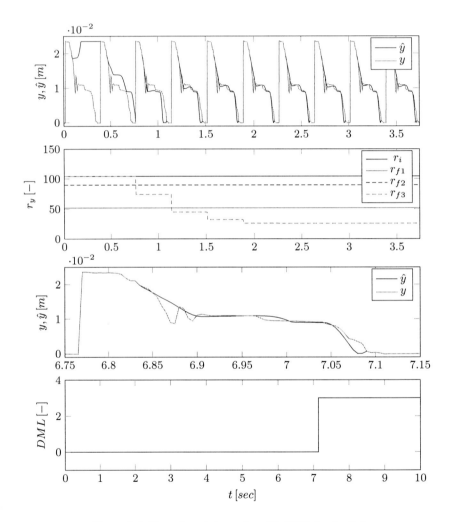

Figure 6.8: Experimental results with DML for fault 3

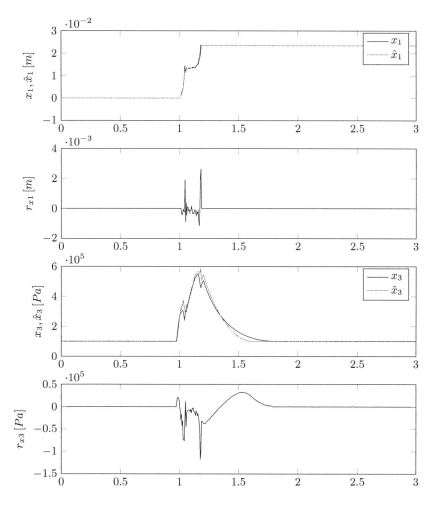

Figure 6.9: Experimental results of the fault free system

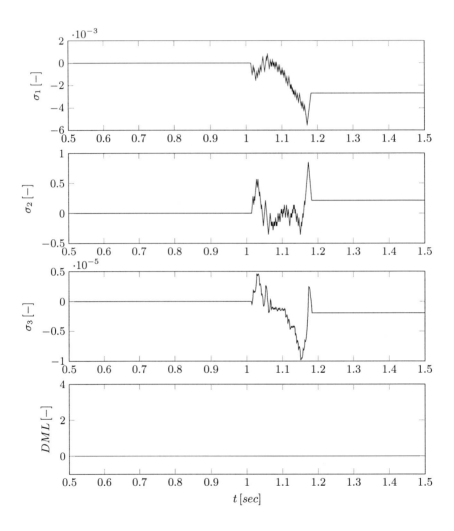

Figure 6.10: Decision making logic for the fault free system

6.3.2 Experiment 2: Leakage in piston seal (fault 1)

A typical example of a leakage in the piston seal is depicted in Figures 6.11 and 6.12. The first subfigure of Figure 6.11 shows very good performance of $\hat{x}_1$ for an inward shifting motion of the actuator having a leakage in the piston seal. Turning to second subfigure of Figure 6.11, which shows the residual $x_1 - \hat{x}_1$, it can be seen that the residual remains less then $1.5 \times 10^{-3}m$, which corresponds to about 7% with respect to the maximum shift of the actuator. As can be seen from the fourth subfigure, the residual between subfigure three's x_3 and $\hat{x}_3$ remains below $1.0 \times 10^5 Pa$ in the shifting phase, which corresponds to about 13% with respect to the maximum chamber pressure of chamber the actuator. As can be seen from first three subfigures of Figure 6.12 σ_1, σ_2 and σ_3 become:

$$\sigma_1 = -1.28 \times 10^{-2} \quad \sigma_2 = 1.27 \quad \sigma_3 = 1.45 \times 10^{-5} \tag{6.5}$$

The output of the decision making logic, which is depicted in subfigure four, correctly outputs value 1 indicating that a leakage in piston seal is present.

6.3.3 Experiment 3: Leakage in piston rod seal (fault 2)

Figures 6.13 and 6.14 depict a typical example of a leakage located in piston rod seal, allowing for a mass flow to exit from chamber A into the surrounding. Again, the first subfigure of Figure 6.13 shows good fit for x_1 and $\hat{x}_1$ for nominal inward shifting motion of the actuator. From the second subfigure of Figure 6.13, which shows the residual $x_1 - \hat{x}_1$, it can be seen that the residual remains less then $1.4 \times 10^{-3}m$, which corresponds to about 6% with respect to the maximum shift of the actuator. As can be seen from the fourth subfigure, the residual x_3 and $\hat{x}_3$ remains below $9 \times 10^4 Pa$ in the shifting phase, which corresponds to about 16% with respect to the maximum chamber pressure of chamber the actuator. Figure 6.14 shows in the first three subfigures σ_1, σ_2 and σ_3 which are:

$$\sigma_1 = -1.50 \times 10^{-2} \quad \sigma_2 = 1.33 \quad \sigma_3 = -1.99 \times 10^{-5} \tag{6.6}$$

The output of the decision making logic, which is depicted in subfigure four, correctly outputs value 2 indicating that a leakage in piston rod seal is present.

6.3.4 Experiment 4: Leakage in cylinder (fault 3)

In the fourth experiment a leakage in cylinder B is introduced. A typical example of a leakage in cylinder B is shown in Figures 6.15 and 6.16. Again the observer performs well, as can be seen from first subfigure of Figure 6.15.

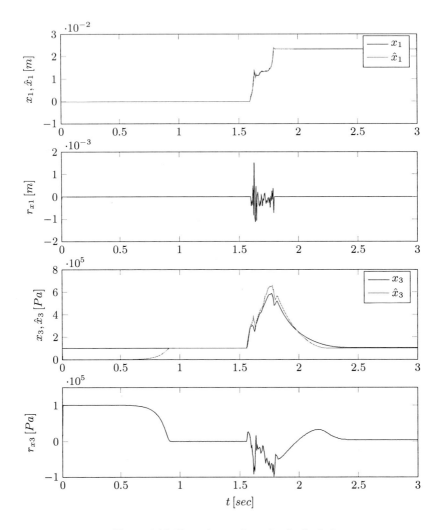

Figure 6.11: Experimental results for fault 1

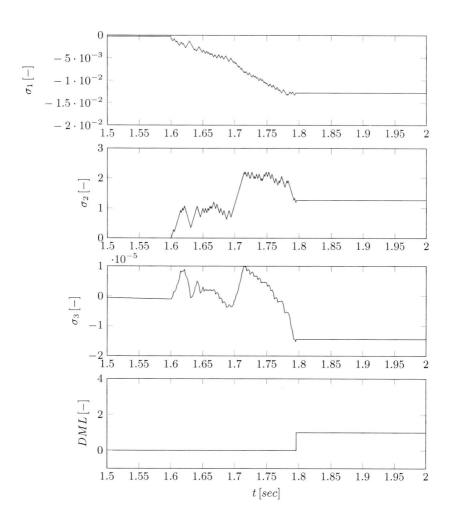

Figure 6.12: Decision making logic for fault 1

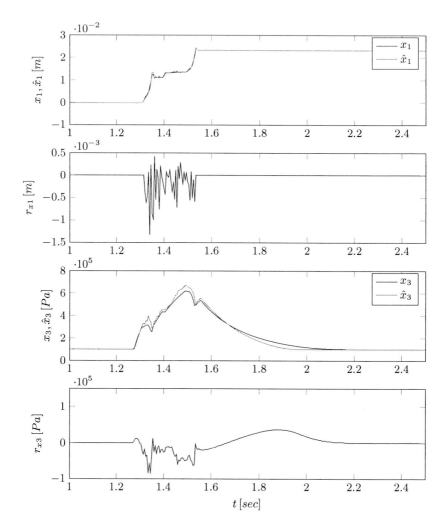

Figure 6.13: Experimental results for fault 2

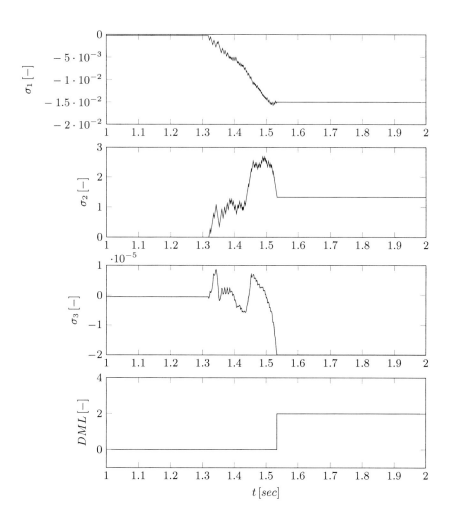

Figure 6.14: Decision making logic for fault 2

Turning to second subfigure of Figure 6.15, which shows the residual $x_1 - \hat{x}_1$, it can be seen that the residual remains less then $3.6 \times 10^{-3}m$, which corresponds to about 15% with respect to the maximum shift of the actuator. As can be seen from the fourth subfigure, the residual $x_3 - \hat{x}_3$ remains below $5.0 \times 10^4 Pa$ in the shifting phase , which corresponds to about 10% with respect to the maximum chamber pressure of chamber the actuator. Note from the third subfigure of 6.15 that the chamber pressure is about 1.5×10^5 Pa lower than that of the prior experiments. This is because of the outward shifting motion of the actuator, wherein chamber B is pressurized. The observer still observes pressure x_3 of chamber A but now as the pressure counteracting the shifting motion of the actuator. As can be seen from first three subfigures of Figure 6.16 σ_1, σ_2 and σ_3 become:

$$\sigma_1 = 1.81 \times 10^{-2} \quad \sigma_2 = -7.04 \times 10^{-1} \quad \sigma_3 = 1.26 \times 10^{-5} \qquad (6.7)$$

As a leakage in piston chamber B is present, the decision making logic output depicted in subfigure four of 6.16 correctly outputs value 3.

6.4 Conclusion

The results show that both diagnosis methods are capable of isolating each of the leakage faults f_1, f_2 and f_3 independently without requiring any chamber pressure sensors. The sliding mode observer scheme is especially convenient due to its simplicity. However, since the observer is designed to maintain a sliding motion in the presence of a fault, gain adaption techniques cannot be applied. Therefore, the sliding mode observer scheme is not capable of determining a fault magnitude. Fault magnitude can be deduced using the parameter based approach, however at a greater computational expense.

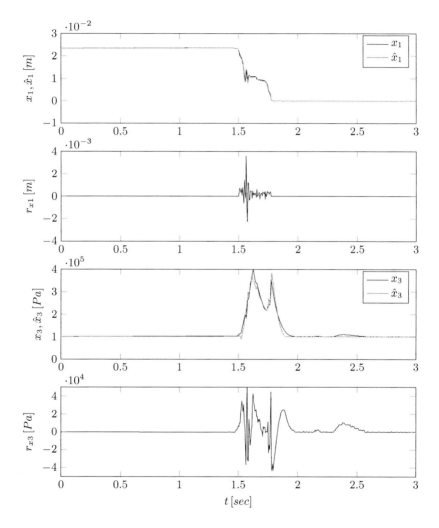

Figure 6.15: Experimental results for fault 3

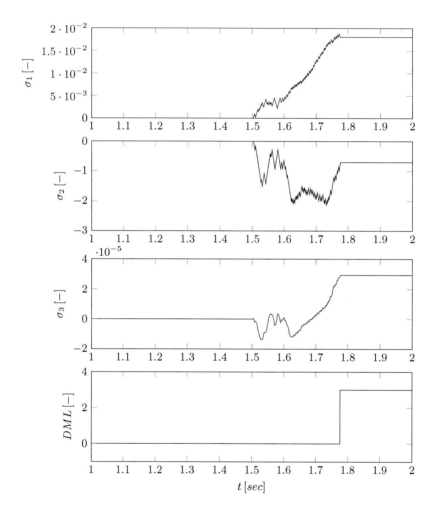

Figure 6.16: Decision making logic for fault 3

7 Summary and outlook

In this thesis a new concept for a distributed diagnosis with signal-based on-board data acquisition and model-based off-board diagnosis is introduced and exemplarily implemented and verified for an electro-pneumatic shift actuator comprised by an automated manual transmission typically used for heavy commercial vehicles. The need of this concept results from the need to perform an effective diagnosis under practical restrictions, namely a limited access to a commercial vehicle's ECUs and a limited numbers available vehicle-sensors. It is shown in this work that an effective diagnosis under these practical restrictions is feasible.

After a short introduction into the field of automotive diagnostics, the new diagnosis concept is presented in Chapter 2. All relevant components of the concept's overall diagnosis chain, namely symptoms-based fault detection unit, ring buffer, compression algorithm, data transfer connection, decompression algorithm, model-based fault isolation unit and fault visualization are explained in detail.

In Chapter 3 a detailed physical model of the electro-pneumatic shift actuator for an automated manual transmission is developed. The electro-pneumatic shift actuator is modeled as a differential pneumatic cylinder with two digital magnetic valves. Furthermore relevant components of the transmission connected to the shift actuator are modeled. The model parameters are identified and validated experimentally using a number of test bench measurements. Result of the identification show very good model fit in terms of chamber pressure and cylinder position. The findings include that the dynamics of the shift actuator is dominated by the large mass of the piston-rod assembly and viscous friction, whereas coulomb friction and external counterforces acting on the piston are negligible and thus can be treated as model uncertainties. Thus, the electro-pneumatic shift actuator at hand is of 'good nature'. At the end of the chapter the model of the electro-pneumatic shift actuator is presented in a state space form. Furthermore introduced in Chapter 3 is the construction of three relevant faults, namely a leakage in the first and second chamber respectively and a cross-chamber leakage. The results achieved in this chapter are the foundation for both model-based fault diagnosis schemes discussed in the next chapters.

As one of the two diagnosis schemes that are presented in this thesis, in Chapter 4 a fault diagnosis based on nonlinear parameter identification is introduced. The problem of residual generation using Levenberg-Marquardt's algorithm and fault induction used for this concept are described in detail. Furthermore as a decision

making logic is formulated which takes both the identified parameters and the generated residual into consideration. The decision making logic also serves the purpose
of testing whether or not the identified parameters are physically plausible. The
chapter concludes with an algorithmic convergence test and initial identification
simulations for a fault detection.

In Chapter 5 a novel design procedure for a fault diagnosis scheme based on a
sliding mode observer is presented. In the beginning of the chapter an observability
analysis of the pneumatic actuator is performed. As the model turns out to loose
observability at certain states in the state space, a reformulation of the model is
performed subsequently. Because of the good nature of the pneumatic cylinder a
numerically calculated pressure for the second chamber can be used to establish
observability. Afterwards a main sliding mode observer for the nominal system and
a respective auxiliary sliding mode observer for each of the three faults is designed.
The observer gains of the main sliding mode observer are tuned such that they
accommodate the fault dynamics of auxiliary sliding mode observers so that the
main sliding mode observer maintains sliding even in the presence of a fault. A
set of values indicative of a SMO's sliding terms is constructed and a DML based
on these values is established. The simulation shows that the DML is capable of
isolating the three faults.

In the beginning of Chapter 6 the setup of a realistic test bench and the appropriate parameter settings for both diagnosis schemes are presented. Measurements
obtained from the test bench for the nominal case and all three faults introduced in
Chapter 3 are used as input data for both the diagnosis scheme based on nonlinear
parameter identification as introduced in Chapter 4 and the sliding mode observer
based scheme of Chapter 5. Notably only signals available in a realistic vehicle are
used as input data for the diagnosis schemes, whereas the additional signals available from the test bench - namely measured chamber pressures - were only used for
verification. The results show that both diagnosis schemes are capable of isolating
each of the respective faults from typical measurements assuming that only a specific fault occurs at a time. Advantageously, the fault parameters of the nonlinear
parameter identification scheme can be used not only to isolate but to determine the
magnitude of a respective fault. As the sliding mode observer is based on constant
gains a fault magnitude cannot be determined. However, this scheme requires less
computational resources.

As part of future work both diagnosis schemes shall be further tested and verified
in terms of robustness regarding the spread of physical parameters between several
vehicles. Later the schemes' applicability to other components shall be tested and
verified.

Kurzfassung

Hintergrund und Motivation

Eine hohe Ausfallsicherheit ist ein wichtiges Kriterium beim Nutzfahrzeugkauf und daher ein vorrangiges Entwicklungsziel. Durch ein rechtzeitiges Diagnostizieren von fehlerhaften Komponenten können Ausfälle des Gesamtfahrzeugs verhindert werden. Die hierfür notwendigen, rechenintensiven Diagnose-Algorithmen sind aus Ressourcengründen meistens nicht auf den für die Komponentenansteuerung zugeschnitten Steuergeräten umsetzbar. Desweiteren steht aus Kostengründen im Fahrzeug auch nur eine begrenzte Anzahl von Sensoren zur Verfügung. Typischerweise wäre auch eine Aktualisierung einer 'On-Board'-Diagnosekomponente mit neu identifizierten Fehlerursachen nur im Rahmen einer jährlich durchgeführten Fahrzeuginspektion und damit häufig zu spät möglich. Ein praxistaugliches Diagnosesystem muss die genannten Randbedingungen berücksichtigen. Abbildung 7.1 zeigt, dass Fehler in

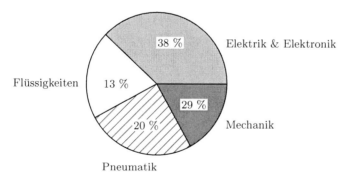

Abbildung 7.1: Gründe für 'Liegenbleiber' schwerer Nutzfahrzeuge [43]

der Fahrzeugelektrik und -elektronik die häufigste Ursache für 'Liegenbleiber' schwerer Nutzfahrzeuge sind. Daher liegt der aktuelle Diagnosefokus bei Nutzfahrzeugen auf der Detektion von Leitungsunterbrechungen, Kurzschlüssen und Sensorausfällen. Die Diagnose pneumatischer Komponenten – auf die sich immerhin ein Fünftel der Liegenbleiber zurückführen lassen – ist bisher vernachlässigt. Da für die ausfallkritische Getriebeautomatisierung elektropneumatische Schaltaktoren zum Einsatz kommen, liegt es nahe diese in die Diagnose einzubeziehen. Die modellbasierte Diagnose ist in vielen technischen Fachrichtungen etabliert und als solche Gegenstand zahlreicher Veröffentlichungen. Aufgrund der Kompressibilität der erforderlichen

Versorgungsluft und der eingangs genannten Sensorknappheit erfordern pneumatische Komponenten einen hohen Modellierungsaufwand. Entsprechende Diagnoseverfahren haben bisher nur bedingt Eingang in die Nutzfahrzeugpraxis gefunden.

In der vorliegenden Arbeit wird ein Fehlerdiagnosekonzept mit einer signalbasierten On-Board Datenerfassung und einer modellbasierten Off-Board Fehlerdiagnose vorgeschlagen. Insbesondere wird untersucht, ob verschiedene Fehlerbilder elektropneumatischer Schaltaktoren unter den eingangs genannten Randbedingungen grundsätzlich diagnostizierbar sind.

Zusammenfassung der einzelnen Kapitel

Nach einer kurzen Einführung in die Thematik der in der Automobilindustrie etablierten Diagnosekonzepte (**Kapitel 1**) werden im **Kapitel 2** zunächst Anforderungen an ein Diagnosekonzept konkretisiert und die Vor- und Nachteile einer funktionalen Diagnoseaufteilung in signalbasierte On-Board Datenerfassung und modellbasierte Off-Board Fehlerdiagnose abgewogen. Im Ergebnis wird eine Diagnosekette mit den On-Board Komponenten symptombasierte Fehlerdetektion ('Trigger'), Ringspeicher, Kompression und den Off-Board Komponenten Dekompression, Fehlerdiagnose, Fehlervisualisierung definiert. Zwischen On-Board Komponenten und Off-Board Komponenten findet eine UDS konforme Kommunikation statt.

Die Diagnosekette ist exemplarisch in Abbildung 7.2 dargestellt und wird im weiteren Verlauf des Kapitels 2 detailliert beschrieben. Über die symptombasierte Fehl-

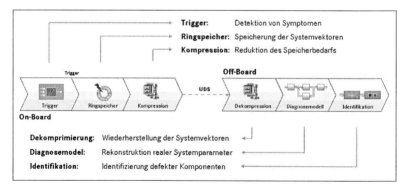

Abbildung 7.2: Diagnosekette mit verteilten Komponenten

erdetektion ist es möglich, die Ansteuer- und Sensorsignale des elektropneumatischen Schaltaktors passiv von einem CAN-Bus des Nutzfahrzeugs auszulesen und eine Schaltzeitverzögerung mittels vorgegebener Schwellwerte zu detektieren. Durch den Ringspeicher kann dabei auf Ansteuer- und Sensorsignale vor einer konkreten Schaltzeitverzögerung zurückgegriffen werden. Mittels einer dem Ringspeicher nachgeschalteten Kompression, werden die Ansteuer- und Sensorsignale des elektropneumatischen Schaltaktors verlustfrei komprimiert. Auf der Off-Board Seite erfolgt eine Dekompression der übermittelten Signale und deren Einspeisung in ein Diagnosemodell. Die modellbasierte Diagnose beruht auf zwei alternativen Paradigmen, zum einen auf der in Kapitel 4 beschriebenen nichtlinearen Parameteridentifikation, zum anderen auf einem Sliding Mode Beobachter, der in Kapitel 5 vorgestellt wird.

Die On-Board Komponenten wurden in Versuchssteuergeräten implementiert und bei Testfahrten (Daimler AG) ausführlich validiert. Auf die dabei konkret ermittelten Messwerte und Fehlerstatistiken kann aus Vertraulichkeitsgründen im Rahmen der vorliegenden Arbeit nicht eingegangen werden. Die Ergebnisse der modellbasierte Off-Board-Diagnose werden ausführlich in Kapitel 6 präsentiert.

Kapitel 3 ist der detaillierten Modellbildung eines für Nutzfahrzeuggetriebe typischen elektropneumatischen Schaltaktors gewidmet und eröffnet mit einem kurzen Stand der Technik zur pneumatischen Modellbildung. Der elektropneumatische Schaltaktor wird in Kapitel 3 als Doppelkammerzylinder modelliert, dessen pneumatischer Teil über thermodynamische Gleichungen hergeleitet wird. Ein digital angesteuertes Magnetventil pro Zylinderkammer befüllt und entlüftet den Doppelkammerzylinder. Die Magnetventile werden als von einem idealen Gas durchströmte Kreisblenden modelliert. Vorteilhafterweise sind die konstruktiven Merkmale des verwendeten Magnetventils exakt bekannt. Die Modellierung des mechanische Teils des Schaltaktors erfolgt unter Berücksichtigung Coulombscher und viskoser Reibung. Der mit dem Kolben des Schaltaktors wechselwirkende Synchronring des Getriebes wird ebenfalls berücksichtigt. Der mechanische und der pneumatische Modellteil des Schaltaktors werden in einem Gesamtmodell zusammengeführt, und die erforderlichen Parameter anhand einer Reihe von Prüfstandmessungen identifiziert.

Durch die Verwendung der identifizierten Parameter zeigt sich bereits eine hohe Modellgüte. Bei einer Simulation mit unterschiedlichen Versorgungsdrücken liegt der Fehler zwischen gemessenem und berechneten Druck unter 10 Prozent, hinsichtlich der Kolbenposition ergibt sich eine Abweichung unter 15 Prozent. Aufgrund seines großen Reservoirs schwankt der Versorgungdruck bei einem Schaltspiel des Schaltaktors um weniger als 1 Prozent. Es hat sich gezeigt, dass die Dynamik des elektropneumatischen Schaltaktors durch seine vergleichsweise große Kolbenmasse sowie durch viskose Reibung dominiert wird. Coulombsche Reibeffekte und auf den Kolben wirkende Getriebegegenkräfte können aufgrund der hohen Betätigungskraft des Schaltaktors von bis zu 700 N vernachlässigt werden. Der betrachtete Schaltaktor ist insofern von gutmütiger Natur. Am Ende des Kapitels 3 wird das Modell des elektropneumatischen Schaltaktors in Zustandsraumdarstellung angegeben. Weiterhin eingeführt werden drei relevante Fehler des Schaltaktors in Form einer Leckage in der ersten und zweiten Zylinderkammer, sowie einer Leckage zwischen den Kammern selbst. Ein typischer elektropneumatischer Schaltaktor für ein Nutzfahrzeuggetriebe

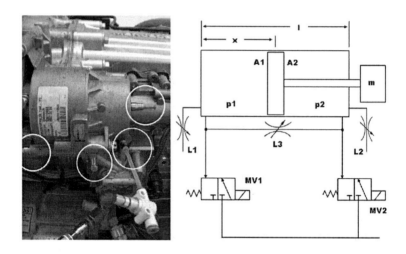

Abbildung 7.3: Elektropneumatischer Schaltaktor mit eingebauter Leckage

ist in der linken Hälfte der Abbildung 7.3 dargestellt. Mit der am Schaltaktor angeordneten Präzisionsdrossel kann eine gewünschte Leckage eingestellt werden. Die rechte Hälfte der Abbildung 7.3 zeigt den Schaltaktor und die drei Leckagen in einer schematischen Darstellung. Das in Kapitel 3 entwickelte Modell dient als Grundlage für die in Kapitel 4 und 5 besprochenen Off-Board Diagnosekonzepte.

Als erstes der beiden Off-Board Diagnosekonzepte wird in **Kapitel 4** die Fehlerdiagnose auf Basis nichtlinearer Parameteridentifikation vorgestellt. Nach einem kurzen Abriss zur Historie der Parameteridentifikation wird die zu lösende Diagnoseaufgabe in ein Minimierungsproblem überführt. Dabei wird zunächst jede der zu diagnostizierenden Leckagen durch einen parasitären Massenstrom beschrieben, der – analog zur Modellbildung der Magnetventile – als ideales Gas eine Kreisblende durchströmt. Diese Durchströmung wird jeweils durch einen parasitären Durchflusskoeffizienten bestimmt, der als Parameter in das Minimierungsproblem Eingang findet. Durch diese Modellbildung wird der funktionale Zusammenhang von Modellparameter und physikalischem Parameter durch eine annähernde Identität dieser Parametes aufgelöst. In Kapitel 4 wird weiterhin eine Entscheidungslogik zur Isolation der drei Leckagefehler definiert. Eine Plausibilisierung der identifizierten Parameter erfolgt dabei zum einen durch vorgegebene Parameterschranken, zum anderen durch eine zu erreichende Mindestverbesserung des Ausgangsfehlers.

Am Ende von Kapitel 4 wird ein Konvergenztest für den Levenberg-Marquardt Algorithmus, der zum Lösen des Minimierungsproblems verwendet wird, durchgeführt.

Zusätzlich wird die Fehlerdiagnose exemplarisch simuliert. Insgesamt zeigt sich, dass der Algorithmus bereits nach 15 Iterationen auf die zweite Nachkommastelle konvergiert. Die Anzahl der Iterationen für die weiteren Untersuchungen wird auf 20 festgelegt, um die Rechenleistung angemessen zu begrenzen. In der exemplarischen Simultation kann ein Leckagefehler erfolgreich isoliert werden.

Kapitel 5 widmet sich dem zweiten der beiden Off-Board Diagnosekonzepte und beginnt mit einer Betrachtung des Standes der Technik zu nichtlinearen Beobachtern im Allgemeinen und Sliding Mode Beobachtern im Speziellen. Anschließend werden einige Beobachtbarkeitskriterien eingeführt und auf das in Kapitel 3 entwickelte Modell in Zustandsraumdarstellung angewandt. Es zeigt sich, dass das System aufgrund der nicht als Eingangsgrößen vorhandenen Kammerdrücke für bestimmte Zustände seine Beobachtbarkeit verliert. Dementsprechend wird eine Transformation des Doppelkammersystems auf ein Einkammersystem vorgenommen und der Druck der zweiten Kammer berechnet. Für das transformierte, leckagefreie System wird in einem ersten Schritt ein Sliding Mode Beobachter als 'Hauptbeobachter' entworfen und ein entsprechender Konvergenznachweis geführt. In einem zweiten Schritt wird für die Leckagefehler, welche wiederum anhand parasitärer Durchflusskoeffizienten modelliert sind, ein entsprechender 'Hilfsbeobachter' zur Bestimmung der Fehlerdynamik entworfen. In einem dritten Schritt wird der für das leckagefreie System entworfene Beobachter so auf die Fehlerdynamik der Hilfsbeobachter eingestellt, dass der Hauptbeobachter auch in Gegenwart eines Leckagefehlers stabil bleibt. Das heißt, dass der Gleitzustand des Hauptbeobachters auch in Gegenwart eines Fehlers erhalten bleibt. Es lässt sich feststellen, dass die schaltenden Teile des Sliding Mode Beobachters für jeden der drei Leckagefehler charakteristisch ausprägt sind. Es wird gezeigt, das diese charakteristische Ausprägung durch neu eingeführte Integral-Kriterien bewertet und für eine Fehlerisolation verwendet werden kann. Die Simulation zeigt, dass durch die Festlegung eines Schwellenwertes für die Integral-Kriterien alle drei Fehler voneinander isolierbar sind.

In **Kapitel 6** wird zunächst der Prüfstandsaufbau und die für die beiden Off-Board Diagnosekonzepte verwendeten Parametereinstellungen beschrieben. Das Herzstück des Prüfstands bildet ein automatisiertes 12-Gang-Schaltgetriebe mit einem pneumatischen Schaltaktor, wie es typischerweise in schweren Nutzfahrzeugen verbaut wird. Das Schaltgetriebe wird durch einen 11-kW Asynchronmotor angetrieben, ein abtriebseitiges Bremsen erfolgt mittels eines 3-kW Asynchronmotors. Die Fahrzeugumgebung ist durch reale Steuergeräte sowie durch eine zweckmäßige Restbussimulation abgebildet. Am pneumatischen Schaltaktor des Getriebes kann mittels einer Präzisionsdrossel eine definierte Leckage eingestellt werden. Der Schaltaktor am Prüfstand weist – im Gegensatz zu einem im Fahrzeug verbauten Schaltaktor – pro Zylinderkammer einen hochgenauen Drucksensor auf. Die somit genau messbaren Kammerdrücke werden allerdings nur für die Verifikation der Diagnosemodelle, nicht für die Diagnose selbst verwendet.

Am Prüfstand gemessene Signalverläufe werden den in Kapitel 4 und 5 vorgestellten Off-Board Diagnosekonzepten zugeführt. Die Ergebnisse sind in Kapitel 6 ausführlich dokumentiert sind. Sie zeigen dass sowohl das identifikationsbasierte als auch das beobachterbasierte Diagnosekonzept in der Lage ist, drei verschiedene am Schaltaktor eingestellte Leckagen anhand der gemessenen Signalverläufe zu isolieren. Dies gilt unter der Bedingung, dass keine zwei Fehler gleichzeitig auftreten. Die vom identifikationsbasierten Diagnosekonzept ermittelten parasitären Durchflusskoeffizienten für die ersten beiden Fehler liegen in einer akzeptablen 30 Prozent-Umgebung der Nominalparameter. Der parasitäre Durchflusskoeffizient für den dritten Fehler ist doppelt so groß wie der Nominalparameter und damit deutlich höher als erwartet. Die beobachteten Kammerdrücke weisen bezüglich der gemessenen Kammerdrücke einen guten Beobachtungsfehler von unter 20 Prozent auf.

Zusammenfassung und Ausblick

In der vorliegenden Arbeit wird ein Fehlerdiagnosekonzept mit einer signalbasierten On-Board Datenerfassung und einer modellbasierten Off-Board Fehlerdiagnose vorgeschlagen. Desweiteren wird exemplarisch gezeigt, dass verschiedene Leckagen eines elektropneumatischen Schaltaktors auch unter Verzicht auf zusätzliche Drucksensoren grundsätzlich diagnostizierbar sind. Das neue Diagnosekonzept ergibt sich aus dem Bedürfnis, das Diagnosespektrum moderner Nutzfahrzeuge im Sinne einer Erhöhung der Ausfallsicherheit zu erweitern, gleichzeitig jedoch mit den Ressourcen der Fahrzeugsteuergeräte sparsam umzugehen. Aus unternehmerischer Sicht liegt ein weiterer Vorteil in der Entkopplung der beteiligten Entwicklungsprozesse und damit der Komplexität der Entwicklungsaufgabe. Bisher konnten Fehlerbilder nur mit Hilfe eines Softwareupdates im Rahmen eines planmäßigen Werkstattaufenthaltes aktualisiert werden. Dieser Aufwand entfällt nun durch den Einsatz eines Standard-Softwaremoduls im Fahrzeug und einer monatlichen Aktualisierung der Diagnosealgorithmen in einer leistungsstarken Off-Board Diagnoseeinheit. Im Rahmen zukünftiger Arbeiten soll die hier vorgestellte Methodik weiter erprobt und später auf weitere Fahrzeugkomponenten übertragen werden.

Bibliography

[1] J. Adamy. *Nichtlineare Regelungen*, chapter 6, pages 287–332. Springer Verlag, 2009.

[2] H. Akaike. A new look at the statistical model identification. *IEEE Transactions on Automatic Control*, 19:716–723, 1974.

[3] E. Alcorta García and P. M. Frank. Deterministic nonlinear observer-based approaches to fault diagnosis: A survey. *Control Engineering Practice*, 5:663–670, 1997.

[4] H. I. Ali, S. Mohd Noor, S. Bashi, and M. Marhaban. A review of pneumatic actuators (modelling and control). *Australian Journal of Basic and Applied Sciences*, 3:440–454, 2009.

[5] T. Assaf and J. B. Dugan. On-board diagnostic expert system via an enhanced fault tree model. In *SAE Technical Paper Series*, 2006.

[6] K. Aström and T. Bohlin. Numerical identification of linear dynamic systems from normal operating records. In *IFAC Symposium on Self-Adaptive Systems*, 1965.

[7] B. Athamena, Z. Houhamdi, and M. Muhairat. Fault detection and isolation in dynamic systems using statistical local approach and hybrid least squares algortihm. *American Journal of Applied Sciences*, 4:977–986, 2007.

[8] J. Barbot, D. Boutat, and T. Floquet. A new observation algorithm for nonlinear systems with unknown inputs. In *Proceedings of the 44th IEEE Conference on Decision and Control*, 2005.

[9] E. J. Barth, J. Zhang, and M. Goldfarb. Sliding mode approach to pwm-controlled pneumatic systems. In *Proceedings of the American Control Conference*, 2002.

[10] R. Beard. *Failure accommodation in linear systems through self-reorganization*. PhD thesis, Massachusetts Institute of Technology, 1973.

[11] D. Bestle and M. Zeitz. Canonical form observer design for non-linear time-variable systems. *International Journal of Control*, 38:419–431, 1983.

[12] B. J. Carr. Practical application of remote diagnostics. In *SAE Technical Paper Series*, 2005.

[13] B. S. E. Chen and Y. T. Chang. A review of system identification in control engineering, signal processing, communication and systems biology. *Journal of Biomechatronics Engineering*, 1:87–109, 2008.

[14] D. Chen and M. Saif. Observer-based strategies for actuator fault detection, isolation and estimation for a certain class on nonlinear systems. *IET Control Theory and Applications*, 1:1672–1680, 2007.

[15] J. Chen and R. Patton. *Robust Model-based Fault Diagnosis for Dynamic Systems*, chapter 2, page 35. Kluwer Academic, 1999.

[16] J. Chen and R. Patton. *Robust Model-based Fault Diagnosis for Dynamic Systems*, chapter 2, page 37. Kluwer Academic, 1999.

[17] L. Chen. *Model-based fault diagnosis and fault-tolerant control for a nonlinear electro-hydraulic system*. PhD thesis, Technical University Kaiserslautern, 2010.

[18] L. Chen and S. Liu. Fault diagnosis and fault-tolerant control for a nonlinear electro-hydraulic system. In *Conference on Control and Fault-Tolerant Systems*, 2010.

[19] L. Chen and S. Liu. Fault diagnosis integrated fault-tolerant control for a nonlinear electro-hydraulic system. In *IEEE Multi-conference on Systems and Control*, 2010.

[20] W. Chen and M. Saif. An actuator fault isolation strategy for linear and nonlinear systems. In *American Control Conference*, 2005.

[21] W. Chen and M. Saif. Actuator fault isolation and estimation for uncertain nonlinear systems. In *Systems, Man and Cybernetics, IEEE International Conference*, 2005.

[22] W. Chen, G. Jia, and M. Saif. Application of sliding mode observers for actuator fault detection and isolation in linear systems. In *IEEE Conference on Control Applications*, 2005.

[23] R. Clark. Istrument fault detection. *IEEE Transactions on Aerospace and Electronics*, 14:356–465, 1978.

[24] Daimler AG. Internal requirements and design specifications, 2007–2010.

[25] S. Drakunov. Sliding-mode observers based on equivalent control method. In *Proceedings of the 31st Conference on Decision and Control*, 1992.

[26] C. Edwards, S. Spurgeon, C. Tan, and N. Patel. *Mathematical Methods for Robust and Nonlinear Control*, chapter 8, pages 221–242. Springer Verlag Berlin-Heidelberg, 2007.

[27] M. Feid. *Systematischer Entwurf signalbasierter Fehlerdiagnosesysteme*. PhD thesis, Technische Universität Kaiserslautern, 2006.

[28] G. Flohr, M. Kokes, A. von Querfurth, and S. Liu. Konzept für eine verteilte funktionsdiagnose mit signalbasierter on-board-datenerfassung und modell-basierter off-board auswertung. In *Mechatronik*, 2009.

[29] T. Floquet and J. Barbot. Super twisting algorithm based step-by-step sliding mode observers for nonlinear systems with unknown inputs. *International Journal of Systems Science*, 38:803–815, 2007.

[30] P. M. Frank. Fault diagnosis in dynamic systems using analytical and knowledge-based redundancy - a survey and some new results. *Automatica*, 26(3):459–474, 1990.

[31] C. Fritsch, J. Lunze, M. Schwaiger, and V. Krebs. Remote diagnosis of discrete-event systems with on-board and off-board components. In *IFAC*, 2006.

[32] A. Girin, P. F., B. X., A. Glumineau, and M. Smaoui. High gain and sliding mode observers for the control of an electropneumatic actuator. In *Proceedings of the 2006 IEEE International Conference on Control Applications*, 2006.

[33] R. Hermann and A. Krener. Nonlinear controllability and observability. *IEEE Transactions on Automatic Control*, 22(5):728–740, 1977.

[34] B. Ho and R. Kalman. Effective construction of linear state-variable models from input/output functions. *Regelungstechnik*, 14:545–548, 1966.

[35] H.-P. Huang, C.-C. Li, and J.-C. Jeng. Multiple multiplicative fault diagnosis for dynamic processes via parameter similarity measures. *Industrial & Engineering Chemistry Research*, 46:4517–4530, 2007.

[36] A. Ilchmanm, O. Sawodny, and S. Trenn. Pneumatic cylinders: modelling and feedback force-control. *International Journal of Control*, 79:650–661, 2006.

[37] *Zwischenringe für Mehrkonus-Synchronisationen.* INA-Schaeffler KG, Automobil Produkt Information API 06 edition.

[38] R. Isermann. Process fault detection based on modeling and estimation methods – a survey. *Automatica*, 22:387–404, 1984.

[39] R. Isermann. *Modellgestützte Überwachung und Fehlerdiagnose für Kraftfahrzeuge*, chapter 18, pages 377–406. Vieweg+Teubner Verlag, 2006.

[40] R. Isermann. *Aktoren*, chapter 10, pages 441–515. Springer Verlag Berlin-Heidelberg, 2008.

[41] M. Jelali. Systematischer beobachterentwurf für nichtlineare systeme. Technical report, Gerhard-Mercator-Universität Duisburg, 1995.

[42] H. Jones. *Failure detection in linear systems.* PhD thesis, Massachusetts Institute of Technology, 1973.

[43] M. Kokes, S. Liu, G. Flohr, and A. von Querfurth. Increasing the fixed first visit quota of an automated manual transmission through a distributed model-based diagnosis approach. In *18. Aachener Kolloquium Fahrzeug- und Motorentechnik*, 2009.

[44] S. Kou, D. Elliot, and T. Tarn. Exponential observers for nonlinear dynamic systems. *Information and Control*, 29:204–216, 1975.

[45] A. Krener and A. Isidori. Linearization by output injection and nonlinear observers. *Systems & Control Letters*, 3:47–52, 1983.

[46] V. Krishnaswami, C. Siviero, F. Carbognani, G. Rizzoni, and V. Utkin. Application of sliding mode observers to automobile powertrain diagnostics. In *Proceedings IEEE International Conference on Control Applications*, 1996.

[47] M. R. Kristensen. Parameter estimation in nonlinear dynamical systems. Master's thesis, Technical University of Denmark, 2004.

[48] I. Lee, J. Kim, D. Lee, and K. Kim. Model-based fault detection and isolation method using art2 neural network. *International Journal of Intelligent Systems*, 18:1087–1100, 2003.

[49] K. Levenberg. A method for the solution of certain problems in least squares. *Quarterly of Applied Mathematics*, 2:164–168, 1944.

[50] C. Li and L. Tao. Observing non-linear time-variable systems through a canonical form observer. *International Journal of Control*, 44:1703–1713, 1986.

[51] A. D. Little. The future of trucks - how technology will change value chain structures, 2004.

[52] J. Liu, B. Jiang, and Y. Zhang. Sliding mode observer-based fault detection and isolation in flight control systems. In *16th IEEE International Conference on Control Applications*, 2007.

[53] L. Ljung. Indentification in nonlinear systems. In *Proceedings of the ICARCV*, 2006.

[54] L. Ljung. Perspectives on system indentification. Technical report, Linköpings universitet, 2008.

[55] K. Madsen, H. Nielsen, and T. O. *Methods for non-linear Least squares Problems*. Technical University of Denmark, 2004.

[56] D. Marquardt. An algorithm for least-squares estimation of nonlinear parameters. *SIAM Journal of Applied Mathematics*, 11:431–441, 1963.

[57] M. Massoumnia. A geometric approach to the synmonogram of failure detection filters. *IEEE Transactions on Automatic Control*, 31:839–846, 1986.

[58] M. Massoumnia, G. Verghese, and A. Willsky. Failure detection and identification. *IEEE Transactions on Automatic Control*, 34:316–321, 1989.

[59] T. Mendoça, H. Alonso, H. Magalhães, and P. Rocha. Multiple strategies for parameter estimation via a hybrid method: a comparative study. Technical report, University of Aveiro, 2005.

[60] H. Németh. *Nonlinear Modelling and Control for a Mechatronic Protection Valve*. PhD thesis, Budapest University of Technology and Economics, 2004.

[61] S. R. Pandian, F. Takemura, Y. Hayakawa, and S. Kawamura. Pressure observer-controller design for pneumatic cylinder actuators. In *IEEE/ASME Transactions on Mechatronics*, volume 7, 2002.

[62] R. Patton, S. Willcox, and J. Winter. A parameter insensitive technique for aircraft sensor fault analysis. *AIAA Journal of Guidance Control Dynamics*, 4: 359–367, 1987.

[63] C. Peris and A. Isidori. A geometric approach to nonlinear fault detection and isolation. *IEEE Transactions on Automatic Control*, 46:853–865, 2001.

[64] G. Qin, H. Zhao, Y. Lei, and A. Ge. A strategy of on-board fault diagnosis of automated mechanical transmission. In *SAE Technical Paper Series*, 2000.

[65] E. Richer and Y. Hurmuzlu. A high performance pneumatic force actuator system part 1 - nonlinear mathematical model. *ASME Journal of Dynamic Systems Measurement and Control*, 122(3):416–425, 2000.

[66] E. Richer and Y. Hurmuzlu. A high performance pneumatic force actuator system part 2 - nonlinear control design. *ASME Journal of Dynamic Systems Measurement and Control*, 122(3):426–434, 2000.

[67] J. J. Rincón-Pasaye and R. Martínez-Guerra. Fault estimation using sliding mode observers. In *Preprints of the 3rd IFAC Symposiun on System, Structure and Control*, 2007.

[68] M. Saif and Y. Xiong. *Fault Diagnosis and Fault Tolerance for Mechatronic Systems*, chapter Sliding Mode Observers and Their Application in Fault Diagnosis, pages 1–54. Springer Verlag, 2003.

[69] A. Schwarte. *Modellbasierte Fehlererkennung und Diagnose des Ansaug- und Abgassystems von Dieselmotoren*. PhD thesis, Technische Universität Darmstadt, 2006.

[70] J. Slotine, J. Hedrick, and E. Misawa. On sliding mode observers for non-linear systems. *ASME*, 109:245–252, 1987.

[71] E. Sobhani-Tehrani and K. Khorasani. *Fault Diagnosis of Nonlinear Systems Using a Hybrid Approach*, chapter 2, pages 21–49. Springer Science+Business Media LLC, 2009.

[72] A. Swarnakar, H. Marquez, and T. Chen. A new scheme on robust observer based control design for nonlinear interconnected systems with application to an industrial utility boiler. In *Proceedings of the American Control Conference ACC '07*, pages 5601–5606, 2007.

[73] F. Thau. Observing the state of non-linear dynamic systems. *International Journal of Control*, 17:471–479, 1973.

[74] K. Veluvolo and Y. Soh. Nonlinear sliding mode observers for state and unknown input estimations. In *Proceedings of the 46th IEEE Conference on Decision and Control*, 2007.

[75] K. Veluvolu, Y. Soh, W. Cao, and Z. Liu. Discrete-time sliding mode observer design for a class on uncertain nonlinear systems. In *Proceedings of the 2006 American Control Conference*, 2006.

[76] K. Veluvolu, Y. Soh, and W. Cao. Robust observer with sliding mode estimation for nonlinear uncertain systems. In *IET Control Theory Appl.*, 2007.

[77] B. Walcott and S. Zak. State observation of nonlinear uncertain dynamical systems. *IEEE Transactions on Automatic Control*, 30:166–170, 1986.

[78] J. Wu, M. Goldfarb, and E. Barth. On the observability of pressure in a pneumatic servo actuator. *Journal of Dynamic Systems, Measurement, and Control*, 126:921–924, 2004.

[79] Q. Wu. *Fault Diagnosis in Nonlinear Systems using Learning and Sliding Mode Approaches with Applications for Satellite Control Systems*. PhD thesis, Simon Fraser University, 2008.

[80] X.-G. Yan and C. Edwards. Robust sliding mode observer-based actuator fault detection and isolation for a class of nonlinear systems. In *44th IEEE Conference on Decision and Control*, 2005.

[81] S. You, M. Krage, and L. Jalics. Overview of remote diagnosis and maintenance for automotive systems. In *SAE Technical Paper Series*, 2005.

[82] A. Zadeh. On the identification problem. *IRE Transactions on Circuit Theory*, 3:277–281, 1956.

[83] Q. Zhang, M. Basseville, and A. Benveniste. Fault detection and isolation in nonlinear dynamic systems: A combined input-output and local approach. Technical report, IRISA, 1997.

[84] J. S. Zypkin. *Grundlagen der informationellen Theorie der Identifikation*. Verlag Technik Berlin, 1987.

Lebenslauf

Ausbildung

05/2007–12/2012   **Technische Universität Kaiserslautern**
     · Externer Doktorand am LRS
10/2010–08/2012   **FernUniversität in Hagen**
     · Recht für Patentanwältinnen und Patentanwälte
10/2003–04/2007   **Technische Universtität Ilmenau**
     · Hauptstudium Mechatronik
02/2003–06/2003   **Universität Nottingham**
     · Produktionsplanung und Management
08/2002–01/2003   **Moskauer Energetisches Institut**
     · Automatisierung elektrischer Antriebe
10/1999–07/2002   **Technische Universität Ilmenau**
     · Grundstudium Mechatronik

Praxiserfahrung

10/2010–01/2013   **Eisenführ, Speiser & Partner**
     · Patentanwaltskandidat
05/2007–09/2010   **Daimler AG**
     · Vorentwicklung Daimler Trucks
01/2006–09/2006   **DaimlerChrysler AG**
     · Entwicklung Mercedes-Benz Lkw
10/2004–04/2005   **DaimlerChrysler AG**
     · Projektleitung Heavy Duty Engine Platform
07/2003–10/2003   **Astro- und Feinwerktechnik GmbH**
     · Produktentwicklung / Versuch
08/2000–10/2000   **Advanced Micro Devices**
     · Materials Analysis Lab
07/1999–09/1999   **Advanced Micro Devices**
     · Materials Analysis Lab